Contents

1 How the Earth Is Built

Key words

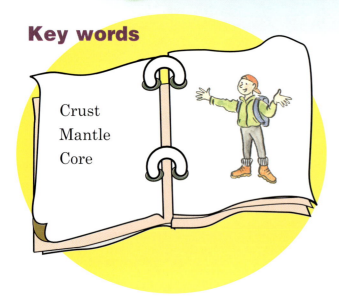

Crust
Mantle
Core

Journey to the Centre of the Earth

If you were to start digging into the ground under your feet, what do you think you would find?

You would dig your way through the soil till you came to solid rock. The solid rock is the **crust** of the earth. It is the thinnest layer.

If you were to continue to dig down, you would find things becoming a little hot and sticky! You would now find yourself in the **mantle**, where it is so hot that the rocks are not solid any more. The temperatures are as high as 4,000°C (degrees Centigrade). Melted rocks flow around, like a hot stew.

If you were to continue to dig further into the centre of the earth you would eventually come to the metal **core** that scientists believe consists of solid rock. Temperatures here are more than 6,000°C (degrees Centigrade).

Just think! All of this is happening underneath your feet.

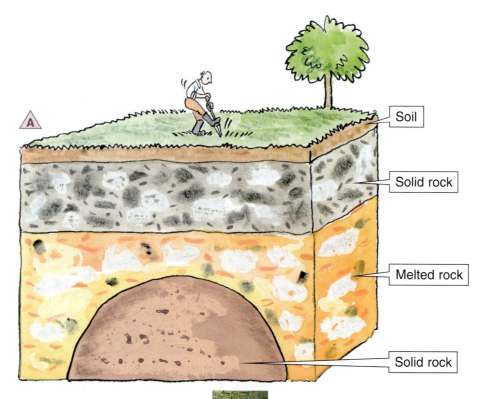

A

Soil

Solid rock

Melted rock

Solid rock

The earth is built like an apple. The **crust** of the earth is like the skin of the apple. The **mantle** is like the flesh of the apple. The **core** is at the centre of both the earth and the apple.

Skin

Core

Flesh

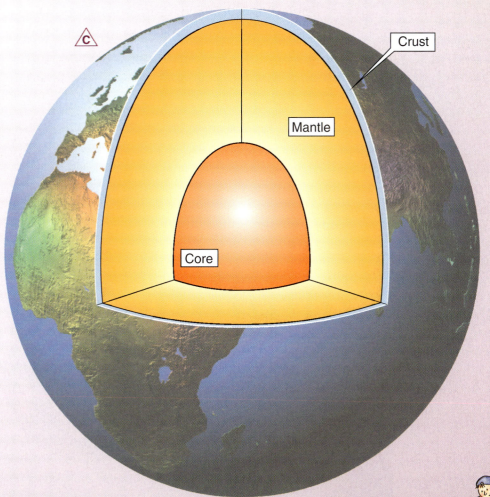

Crust

Mantle

Core

Look at diagram C. It shows how the earth is built.

Key points

- The earth is made up of three layers. These are the **core**, **mantle** and **crust**.
- The **crust**, the outside layer, is the thinnest layer.
- Rocks get hotter and melt as you travel into the earth.

Crust – The Skin of the Earth

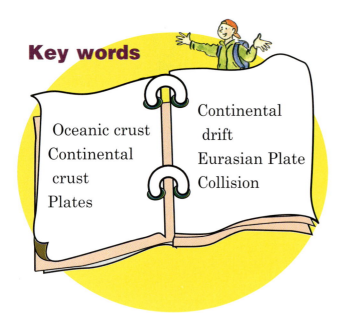

Key words

Oceanic crust
Continental crust
Plates

Continental drift
Eurasian Plate
Collision

Let's look at the crust in more detail.

Continental and oceanic crust

Most of the crust – 75% – lies under the sea. This part of the crust is called the **oceanic crust**. It forms the floor of the ocean. It is made of heavy rock. The rest of the crust appears as land and we call this **continental crust**. It is made of lighter rock.

Plates

The crust is broken up into pieces called **plates**. There are seven major plates. The plates fit together like the pieces of a jigsaw puzzle.

Plate movement

Plates rest on top of the flowing hot rocks in the mantle. The movement of the flowing rocks underneath pulls the plates around. Think of goods moving along the conveyor belt at the supermarket checkout. This is what happens with plates. They are moved around. They may only be moved a few centimetres a year, but over millions of years this makes a huge difference. The continents used to be joined together in a single continent, but they drifted apart. This movement is still happening. This is called **continental drift**. Look at diagram A.

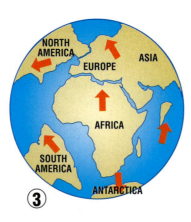

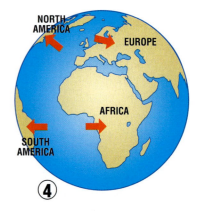

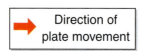
Direction of plate movement

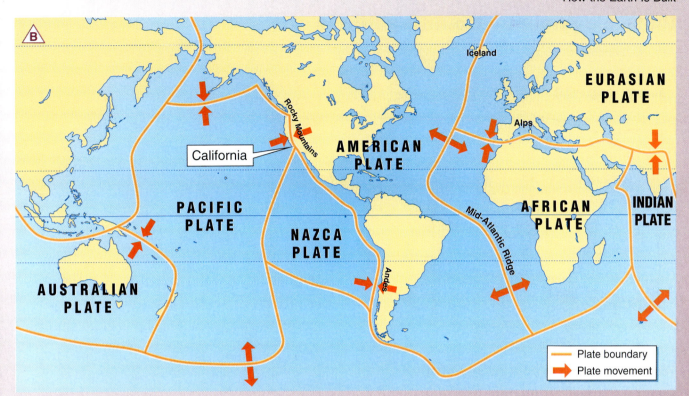

Naming the plates

Geographers like to know the names of places so the plates have all been given names. They are named after the oceans or continents. One plate is called the **Eurasian Plate** (after Europe and Asia). Ireland is on this plate. Look at map Ⓑ that shows this. California is at the edge of two plates – the *Pacific* and the *American*.

Three types of plate movement

- Some plates are moving towards each other, causing **collision**.
- Others are moving apart resulting in separation.
- Others are sliding past each other causing enormous friction and cracking of the crust.

Key points

- The **crust** is the solid rock layer, including the ocean floor and the continents.
- The crust is divided into pieces called **plates**.
- The **plates** are being moved by the flowing hot rocks in the mantle.
- The **plates** move together, apart and alongside each other.

Plate Movement – Colliding Plates

Key words

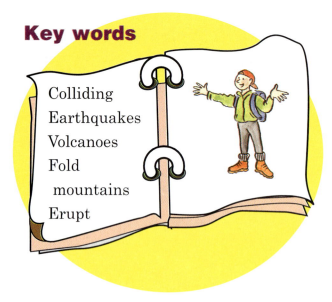

Colliding

Earthquakes

Volcanoes

Fold
  mountains

Erupt

Plates are moved about on top of the flowing hot rocks (magma) in the mantle. In some places these plates are **colliding** together. Let's look in more detail at what happens at a collision zone.

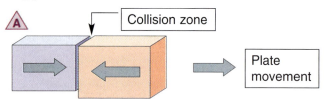

A | Collision zone

Plate movement

Pressure builds up where the two plates are colliding. This can lead to **earthquakes**, **volcanoes** and **fold mountains**.

Earthquakes

When two plates collide the solid rocks of the crust bend and crack under the strain. The ground shakes in an **earthquake**.

Volcanoes

The crust may be destroyed in the collision. It may melt. Hot melted rock may rise up through the crust and **erupt** as a **volcano**.

Fold mountains

Sometimes the edges of crashing plates are buckled up into folds. This action creates **fold mountains**.

Case Study:
West Coast of South America

The west coast of South America shows what happens when two plates collide. This is an interesting example because all three effects are seen here.

When the Nazca Plate (the plate under the Pacific Ocean) pushes against the American Plate, the rocks along the edges are squeezed together. The ground shakes setting off earthquakes. Earthquakes happen everyday along the west coast of South America.

The force of the collision between the plates pushes up the Andes Mountains into great folds.

While the mountains are being pushed up, the Nazca Plate is dipping down under the South American Plate. As it dips down into the mantle, it melts in this hot zone. The melted rock rises up through cracks in the mountains above it. It forces its way through and finally **erupts** or spills out as a volcano.

Look at diagram **C**. It shows what is happening in this collision zone.

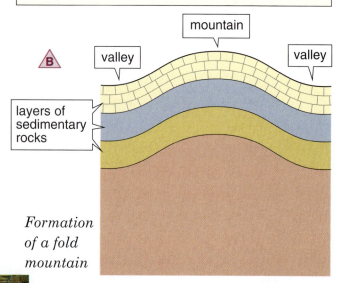

B | valley | mountain | valley

layers of sedimentary rocks

Formation of a fold mountain

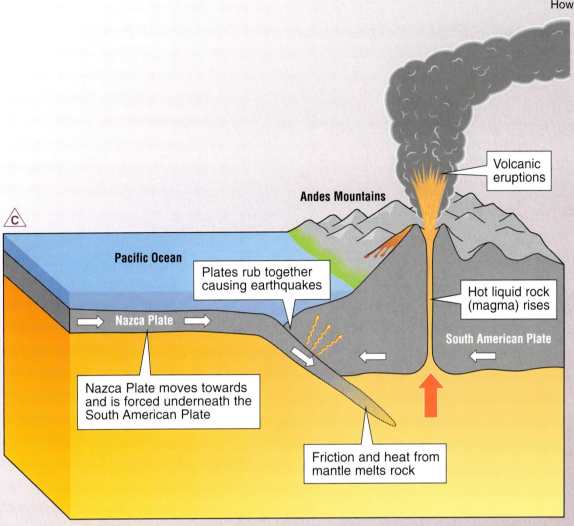

Ⓒ

Volcanic eruptions

Andes Mountains

Pacific Ocean

Plates rub together causing earthquakes

Hot liquid rock (magma) rises

→ **Nazca Plate** →

South American Plate

Nazca Plate moves towards and is forced underneath the South American Plate

Friction and heat from mantle melts rock

Ⓓ

Case Study: India in Collision with Asia

Millions of years ago India was locked into the side of Africa (see figure Ⓐ, page 4). Plate movement has pulled it northwards where it is now colliding with Asia. The northern edge of India and the southern end of China are being squeezed up in the collision. This has made the fold mountains called the Himalayas. These are the world's highest mountains. Mount Everest is in this range of mountains and rises to over 8,848 metres (29,029 feet).

Earthquakes are common in this collision zone.

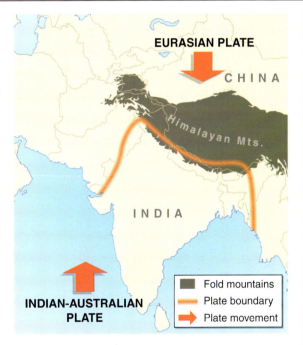

EURASIAN PLATE

C H I N A

Himalayan Mts.

I N D I A

INDIAN-AUSTRALIAN PLATE

Fold mountains
Plate boundary
Plate movement

Key points

● **Earthquakes**, **volcanoes** and **fold mountains** are a result of plate collision.

7

Plate Movement – Plate Separation and Plates Sliding

Key words

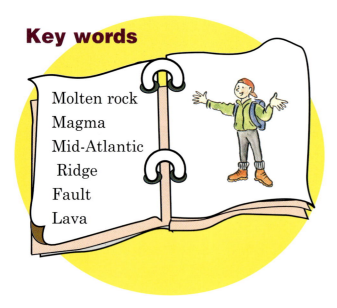

Molten rock
Magma
Mid-Atlantic
 Ridge
Fault
Lava

Plate separation

When two plates move apart, this leaves a gap. Hot **molten rock** called **magma** rises up from under the crust to fill the gap. It spills out onto the surface of the earth. It is now called **lava** and it hardens to form new crust.

Mid-Atlantic Ridge and volcanic islands

The **Mid-Atlantic Ridge** marks the place where plates are separating. It is a line of volcanoes running from north to south under the Atlantic Ocean. In places, this ridge rises above the sea to make islands e.g. Iceland.

Plates sliding alongside each other

When two plates slide past each other, there is a great deal of friction between them. This makes the movement jerky. The plates may become jammed together and then suddenly slip. This causes an earthquake. This is common along the state of California, where two plates are sliding past each other.

The **San Andreas Fault** in California is a long crack on the earth's surface. It shows where two plates are sliding alongside each other.

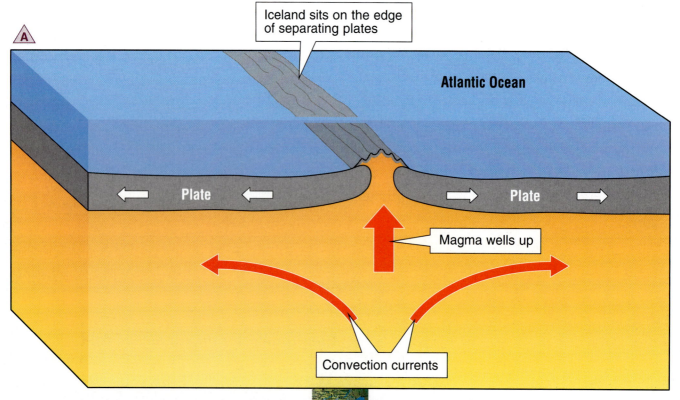

Iceland sits on the edge of separating plates

Atlantic Ocean

Plate

Plate

Magma wells up

Convection currents

San Andreas Fault

Case Study: Iceland

Think of Iceland as a country that is being stretched from side to side with a big split down the middle. The country is being pulled in opposite directions on separating plates. **Lava** is pouring out to fill the gap. In 1963/64, lava piled up to make a new island called Surtsey.

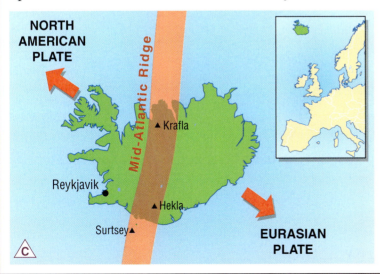

Key points

- **Magma** spills up from the mantle to fill the gap when two plates move apart.

- **Lava** is the **molten rock** that pours out to make a volcano.

- Earthquakes happen where plates slide past each other, and where plates are separating.

- The **Mid-Atlantic Ridge** is a line of volcanoes where two plates are separating.

Volcanoes

Key words

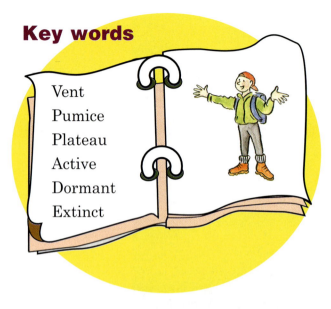

Vent
Pumice
Plateau
Active
Dormant
Extinct

Mount Vesuvius

Mount Vesuvius is a famous volcano south of the city of Naples in Italy. When Mount Vesuvius erupted in AD 79 the nearby city of Pompeii was buried under hot ash. Many of the people were poisoned by the gases from the volcano.

Mount Vesuvius

Volcanoes – cone-shaped mountains

When two plates move apart a gap opens up. Magma (molten rock) rises up from the mantle to fill the gap. This action makes volcanoes.

If the gap is a narrow opening or hole in the crust, this is called a **vent**. Hot molten rock erupts from this vent. After many eruptions, it builds up into a cone-shaped mountain.

Look at diagram A that shows the different parts of a typical cone-shaped volcano.

Some volcanoes erupt very violently. Hot ash, cinders and pumice are blasted high into the air. **Pumice** is light lava with lots of air spaces in it. You may have heard of a pumice stone for removing hard skin from feet! These materials fall back down and cool. Poisonous gases, some with strong smells, also escape.

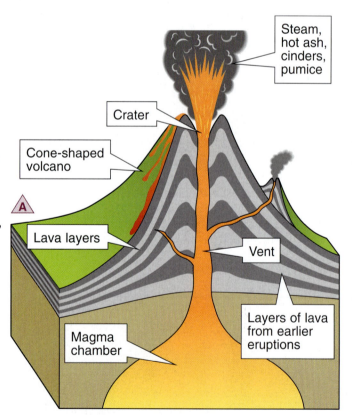

Plateau

When the lava spills out through a long crack rather than a vent, it forms a **plateau**. This is a flat-topped mountain. The Antrim Plateau in Ireland was formed in this way.

Distribution of volcanoes

Look at the map where you would find volcanoes. They lie around the edge of the plates.

There are lots of volcanoes in the world but they are not all **active** or 'live'. Some are described as **dormant** or sleeping because they have not erupted in a long time. Others are described as **extinct**. They have died out. It is unlikely that they will ever erupt again.

Active: Mt St Helens in the USA

Dormant: Mt Vesuvius in Italy

Extinct: Slemish Mountain in Co. Antrim

Key points

- Volcanic activity happens at the edges of plates.
- Lava erupts through a **vent** or a crack.
- The lava builds up to make a cone-shaped mountain called a volcano.
- There are three types of volcanoes: **active, dormant** and **extinct.**

Humans and Volcanoes

Key words

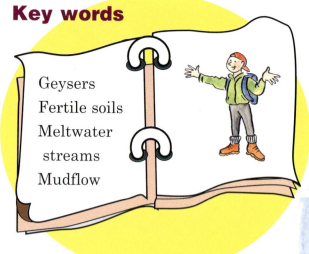

Geysers
Fertile soils
Meltwater
   streams
Mudflow

Mining

Valuable minerals such as gold and diamonds are found near volcanoes. This is the case in South Africa.

Look at the picture and spot the benefits of volcanoes

Although dangerous, volcanic eruptions have many benefits (good effects).

Good effects of volcanoes

Electricity

Water, heated by hot rocks below the surface, makes steam which can be used to make electricity. Sometimes the hot water shoots to the surface in **geysers**. In Iceland, there are many open-air swimming pools where the water has been warmed naturally by the hot rocks underneath.

Farming

Volcanic rocks, when broken down by the weather, make very **fertile soils**. The soils on the slopes of Mt Etna are used to grow crops like tomatoes, grapes and olives.

Tourism

People like to visit areas famous for volcanoes. One example is Pompeii where the town was buried under ash that was thrown up from Mt Vesuvius.

Mudflow buries Almero, Colombia

Bad Effects of Volcanoes

Case Study: Nevada del Ruiz, Colombia

Colombia, a country in South America, is in a place where two plates are colliding. The three effects of collision can be seen here. The Andes Mountains have been folded up here, earthquakes frequently rock the land and volcanoes erupt.

On 13 November 1984, the volcano of Nevada del Ruiz erupted violently. Hot ash and burning rocks were thrown over 8,000 metres into the air. The snow and ice on the sides of this very high mountain melted and flowed down into the valleys below. When the hot ash fell down it mixed with the **meltwater streams**. It then made a huge river of mud that flowed towards the town of Almero. Over 22,000 people were buried in this **mudflow**.

Be prepared!
Forecasting a volcanic eruption

In December 2000, scientists in Mexico City predicted an eruption from the nearby volcano, Popocatepetl. They used instruments that showed that magma had moved into the area beneath the volcano. As it came closer to the surface the magma released gases. This movement of magma also produced small earthquakes and vibrations. Tens of thousands of people were evacuated (moved out of the area). As a result of the prediction many lives were saved.

Almero

Key points

- Volcanoes produce **fertile soils**, minerals, hot water (electricity) and valuable rocks.
- Volcanic sites are interesting places for tourists to visit.
- Volcanoes can kill people, destroy property and land and cause dangerous **mudflows**.
- The **predictions** of volcanic eruptions have improved greatly.

Earthquakes – Shaking of the Crust

Key words

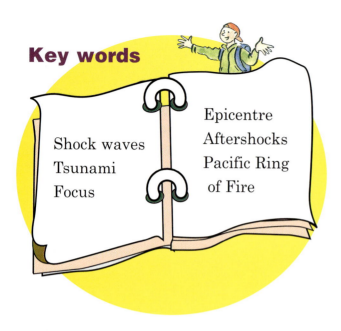

Shock waves
Tsunami
Focus

Epicentre
Aftershocks
Pacific Ring
of Fire

Earthquakes happen at the plates' edges

Earthquakes (also called tremors) happen where plates meet (plates' edges), because the rocks are grinding and scraping together. This causes tension and stress to build up. **Shock waves** then pass through the crust right up to the surface causing an earthquake.

During an earthquake the ground shakes under your feet. Buildings can sway from side to side and sometimes they topple down. Whole cities can be destroyed in a matter of seconds. Many people die instantly, buried alive under heaps of brick or soil.

Roads collapse, railway lines crack and separate. Fires spread when gas pipes burst. Sewage and water pipes split.

Even the sea is shaken about. Giant waves called **tsunamis** are stirred up and can flood the coasts leading to further loss of life and property.

Tsunami Disaster in the Indian Ocean

More than 290,000 people died and over 10 million were left homeless following a tsunami in the Indian Ocean on December 26th 2004. A huge earthquake, measuring 9.0 on the Richter scale, triggered the tsunami.

Sudden movement of the Indian Plate under the Burma Plate caused the crust to jerk upwards pushing up billions of tons of water. This set off the giant waves of the tsunami.

Within a few hours, the 10 m (35 ft) waves had destroyed villages, tourist resorts, water supplies and soils in more than ten countries. There was no early warning system in place. Tourists and locals had no time to move to higher ground.

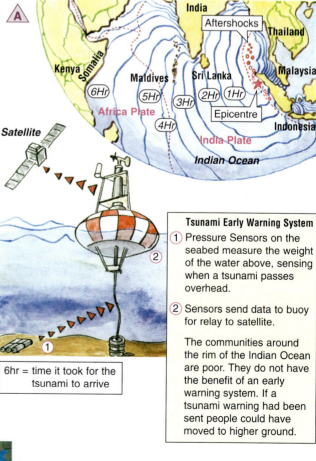

6hr = time it took for the tsunami to arrive

Tsunami Early Warning System

1 Pressure Sensors on the seabed measure the weight of the water above, sensing when a tsunami passes overhead.

2 Sensors send data to buoy for relay to satellite.

The communities around the rim of the Indian Ocean are poor. They do not have the benefit of an early warning system. If a tsunami warning had been sent people could have moved to higher ground.

The focus of the earthquake

Geographers use the word **focus** to describe the place under the ground where the plates are colliding. The focus could be several kilometres deep down in the crust.

The epicentre of an earthquake

You would not like to be standing near the **epicentre** when an earthquake strikes. This is the place where the earthquake is strongest and does most damage. The epicentre is the place on the surface right above the focus.

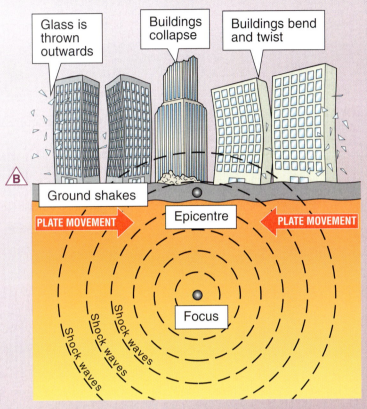

Glass is thrown outwards

Buildings collapse

Buildings bend and twist

Ground shakes

PLATE MOVEMENT

Epicentre

PLATE MOVEMENT

Shock waves

Focus

B

Aftershocks

Smaller shocks called **aftershocks** can continue for many days after a big earthquake. People would be afraid to go back to their houses – that is if their houses were still standing!

Earthquakes in Ireland

Could a serious earthquake ever happen in Ireland? This is not likely. Ireland is not near the edge of a plate and this is where most earthquakes happen. Check this out on the world map showing plates.

If you lived in Japan it would be a different story! It is in the zone called the **Pacific Ring of Fire**, so called because it marks the edge of plates around the Pacific Ocean. It is a place where earthquakes and volcanoes are common.

Key points

- Earthquakes happen at the edges of moving plates.
- Earthquakes begin deep within the crust at the **focus**.
- The point on the earth's surface where the damage is greatest is called the **epicentre**.

Measuring the Strength of an Earthquake

Key words

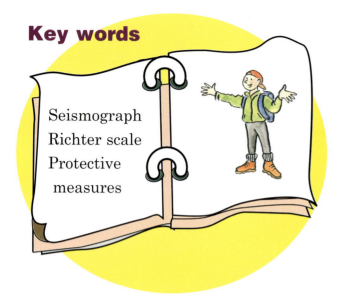

Seismograph
Richter scale
Protective measures

Seismograph

A **seismograph** is used to measure how strong an earthquake is. A reading is traced out on a graph.

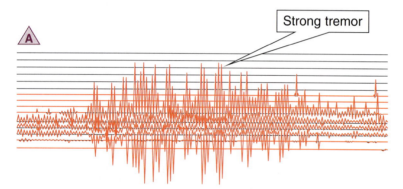

Strong tremor

A

A seismograph recording

Richter scale

The strength of the earthquake is ranked on a scale called a **Richter scale** – 6.5 would be quite strong and would cause a lot of damage while 3.5 might only shake the light bulb! Much depends on where the earthquake happens.

Asian tsunami 2004	9
Assam, India 1950	8.7
San Francisco 1906	8.2
Gujarat, India 2001	7.9
San Francisco 1989	6.9
Italy 1997	5.7

B

Comparison: Earthquakes in poor and rich countries

The damage caused by an earthquake is often far worse in a poor country than in a rich country.

In poor countries:
- Buildings are made of inferior materials which collapse easily during an earthquake.
- There is little money to pay for repairs to roads and houses.
- The people often live in overcrowded conditions so the death toll is high.

In rich countries:
- Buildings are made of better materials which are less likely to collapse.
- There is more money available to pay for repairs to roads and houses.
- The people are more educated and are better able to deal with earthquake emergencies.

Case Study: Kobe, Japan 1995

Japan has 7,000 to 8,000 earthquakes each year. This is more than any other country.

Location: Kobe
Strength: 7.2
Death toll: 6,000
Injured: 35,000
Homeless: 300,000
Cost of damage: US$ 100 billion.

Earthquake drill in Japan

Tall buildings can withstand earthquakes if properly built

Living with earthquakes – protective measures

Nobody can stop an earthquake from happening, but there are a few things that could be done which might mean there is less damage.

- Buildings could be built to sway a bit without falling down, such as in California, USA.
- Earthquake drills (like fire drills) could be practised (they do this in Japan).
- Animals are good at sensing tremors in the ground so people could watch them for early signs of disaster (they do this in China).

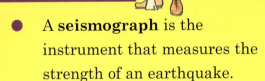

Key points

- A **seismograph** is the instrument that measures the strength of an earthquake.
- **The Richter scale** allows us to compare one earthquake with another.
- Figures can hide the real damage of earthquakes.
- **Protective measures** may lessen the damage of an earthquake.

17

Fold Mountains

Key words

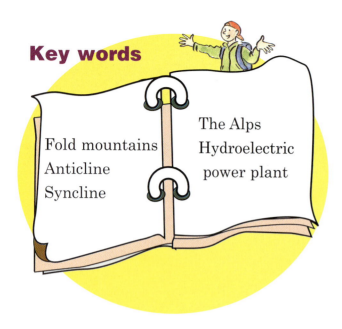

Fold mountains
Anticline
Syncline

The Alps
Hydroelectric
power plant

Fold mountains in Munster, Ireland

Many of the mountains in Ireland are fold mountains. The highest mountains are in Munster. Here, you find the Macgillycuddy Reeks. They include Ireland's highest mountain, Carrauntohill, at over 1,000 metres. These fold mountains were formed about 250 million years ago. They are called Armorican Fold mountains. They were formed at the same time as the Armorican mountains of north-west France.

Fold mountains are formed when two plates crash or collide together. The rocks that are close to the edge of the plates are squeezed up when the plates collide. They form fold mountains.

The top of the fold is called the upfold or **anticline**. The dip between the folds is called the downfold or **syncline**. (Think of pushing a piece of cloth up into folds.)

Fold mountains in Munster

*Many of the world's largest mountains are also fold mountains: **the Alps**, the Himalayas and the Andes*

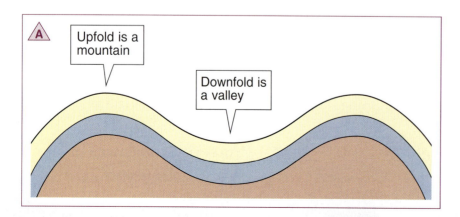

A

Upfold is a mountain

Downfold is a valley

C

Fold mountains and their usefulness to people

The Alps are home to over 11 million people. This mountain chain lies along many countries in southern Europe. They attract people to live and to holiday there because:

- The slopes of the mountains are excellent for growing trees.

- The beautiful scenery found there attracts many tourists all year round such as walkers and skiers. This means more money, jobs and services for the area.

- The steep slopes and the wide valleys are ideal for building **hydroelectric power plants**. Rivers flow very quickly down the slopes and so their energy can be trapped and used in homes and factories.

- The valleys are sheltered and so dairy (milk) cattle do well.

Key points

- **Fold mountains** are found at the edges of colliding plates.
- Many of the world's famous mountains are **fold mountains**.
- Fold mountains have a valley called a **syncline** and a mountain called an **anticline**.

REVISION EXERCISES

Write the answers in your copybook.

1   Look at the map of plates on the surface of the earth on page 5.
    Ireland is on:
    ● The Eurasian Plate
    ● The South American Plate
    ● The African Plate
    ● The Antarctic Plate

2   Earthquakes often occur:
    ● In Ireland
    ● At the centre of the plates
    ● At the edge of the plates
    ● In tundra zones

3   The Pacific Ring of Fire shows:
    ● The location of many earthquakes and volcanoes
    ● An area subject to bush fire
    ● An area of hot springs
    ● An area with a hot desert type climate

4   In your copybook write the correct answer for each of the statements
    below.
    (a) The mountains of Kerry are **volcanic/fold mountains**.
    (b) The centre of the earth is called the **mantle/core**.
    (c) The mountain built by a volcanic eruption is called a **cone/a vent**.
    (d) An earthquake's strength is measured by a **seismograph/a
        Richter scale**.

5   Draw out the diagram of a volcano given below. Label the vent, cone
    and crater.

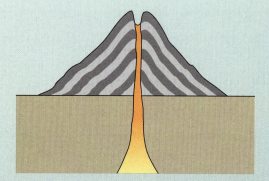

6   Earthquakes and volcanoes are found close to plate edges.
    Give two examples that support this statement.

7   (a) Name one feature associated with the following:
        (i)   separating plates
        (ii) colliding (moving towards each other) plates.

    (b) In the case of each feature that you have chosen:
        (i)   draw a labelled diagram
        (ii) give an example of where you can see this feature
        (iii) write at least three sentences explaining how each feature
              was formed.

8   Explain how magma rising to the surface of the crust can cause
    mudflows. Use an example e.g. Colombia in South America.

9   In your copybook match each letter in column X with the number of
    its pair in column Y.

X	Y	ANSWER
A Focus	1 Fold mountains	A =
B Crater	2 Earthquakes	B =
C Syncline	3 Crust of the earth	C =
D Plates	4 Volcano	D =

10 Imagine you are a TV reporter at the site of an earthquake. Write
    out your report in full. Use details to do with this subject such as:
    Richter scale, epicentre, house collapse, power cuts, the cracking
    open of roads.

11 List four benefits of living close to an active volcano.

12 Fold mountain landscapes attract many people. Name one area of
    Europe that has fold mountains. Give two reasons why people like
    living in these areas.

13 Name two ways used to predict earthquakes.

14 Name two types of material that are blasted out by volcanoes.

2 Rocks

Key words

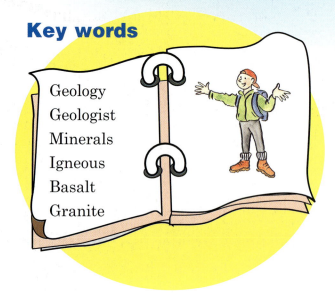

Geology
Geologist
Minerals
Igneous
Basalt
Granite

Geology

The study of rocks is called **geology**. A **geologist** studies rocks. They find out how rocks were made. Knowing these things helps us to understand landscapes.

Minerals in rocks

A walk along a stony beach will show you that there are many different types of rocks. Some rocks are pale in colour, some are dark and some are speckled with shiny pieces in them. Some rocks feel smooth to touch, while others feel rough. If you tried to scratch some rocks with metal you might find it impossible, while other rocks would be quite soft and powdery. These differences happen because different rocks are made up of different **minerals**.

The **minerals** are the ingredients of rocks – just like different ingredients in cakes. Some minerals are precious because they are rare e.g. diamonds, emeralds and gold.

The origin of rocks

Geologists group rocks according to their origin i.e. how they began. There are three groups:

- **Igneous** rocks
- Sedimentary rocks
- Metamorphic rocks.

Igneous rocks

Igneous rocks are a group of rocks that are formed when molten rock cools and hardens. We will look at two examples of **igneous** rock – **basalt** and **granite**.

The Giant's Causeway in Co. Antrim

Basalt rock is used in the making of roads

Basalt

Basalt rock is smooth and black in colour. It is a hard rock. You would find it difficult to scratch it with a nail. It is used to build roads. It is found in Co. Antrim.

Basalt began its life in the layer below the crust. Hot molten rock spilled up through a volcano and flowed out onto the surface of the earth. It flowed quickly across the ground and cooled rapidly. It covered the ocean floors of the world. Then it cooled quickly leaving the minerals to harden into tiny crystals. You can see these crystals under a microscope.

> **BASALT** = black and smooth found in Antrim when lava cooled on top.

Granite

Granite is a speckled (spotty) rock and can be grey, pink or even green in colour. It is a very hard rock and rough to the touch. It is used in buildings. It is found in counties Wicklow, Donegal and Galway.

Granite, like basalt, began its life in the layer of liquid rock below the crust. As it flowed up to the surface it began to cool down. It cooled slowly and *hardened before*

Granite rock is used as a building stone

it got to the top. Slow cooling means that the minerals cool into large crystals. You can see these clearly and feel them too. Its speckled appearance shows you that it is made up of a few different minerals. These minerals are quartz, feldspar and mica (the shiny bits).

> **GRANITE** = grey, rough and cooled in the earth.

Key points

- There are three groups of rocks: **igneous**, metamorphic and sedimentary.
- Rocks are made of **minerals**.
- **Igneous** rocks were formed when lava cooled and hardened.
- **Basalt** is cooled *on* the surface.
- **Granite** cools *underneath* the ground.

Sedimentary Rocks

Key words

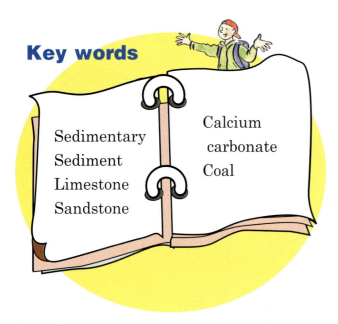

Sedimentary
Sediment
Limestone
Sandstone

Calcium
carbonate
Coal

Sedimentary rocks are made from bits of other rocks, plants or creatures which are washed into the sea and remain on the sea floor for many millions of years. Over this time the **sediments** are gradually (compressed) pressed together under the weight of the sea and are cemented to form a new rock. Let's look at two rocks formed in this way – **limestone** and **sandstone**.

We are all going to be **rock stars**

The word **sediment** means the remains or dregs of something. Think of the sediment of sugar remaining in your teacup if you do not stir it properly.

Fossil in limestone rock

Limestone

Limestone was formed when the remains of dead sea creatures drifted down to the sea floor. These bones and shells built up on the bed of the sea over a long time. Over millions of years sediments pressed down on each other. This resulted in the making of a solid rock called limestone. Sometimes you will be able to see *fossils* of dead sea creatures in a lump of limestone.

Limestone is a grey rock. It is smooth to touch. If you put a spot of *hydrochloric* acid on limestone you would see a fizzing reaction. This is because limestone contains the mineral **calcium carbonate** (from bones and shells).

Limestone is the most common rock in Ireland. It is found all over the Central Plain of Ireland. The rock is covered in soil and bog in some places, but in Co. Clare there is no soil cover. The forces of weather and erosion have removed the covering of soil.

Limestone is very useful. It can be used in the making of steel. It can be ground down and used as a fertiliser on the soil. It is also used as a building stone.

Sandstone

Sandstone, as its name tells you, is made from grains of sand. As with limestone, the grains of sand were buried under the sea. They were pressed down and cemented together to form solid rock.

Sandstone varies in colour from yellow to dark red. It is smooth to touch. It is possible to see the tiny grains of sand with a microscope. It is a hard rock so it usually forms high mountains. A sedimentary rock called old red sandstone is found in the mountains of Cork and Kerry.

D

Coal

Coal is also a sedimentary rock. It is formed from vegetation that has been buried under the sea. Coal is used as a fuel. It is hard to imagine that coal began its life as a forest!

E

Key points

- **Sedimentary** rocks are formed from the remains of other rocks, plants and creatures.
- The **sediments** (remains) are compressed under water and cemented together.
- **Limestone, sandstone** and **coal** are sedimentary rocks.

Sandstone is used as a building stone

Metamorphic Rocks

Key words

Metamorphic
Marble
Quartzite

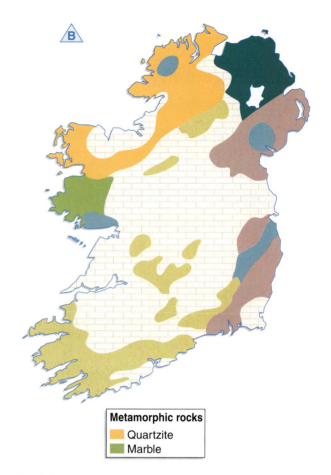

Metamorphic rocks are rocks that are changed due to heat or pressure. **Marble** and **quartzite** are examples of metamorphic rocks.

A surge of magma (boiling rocks from below the crust) can rise and heat limestone to the point where the limestone becomes extremely hot. Limestone later cools to form a new rock called marble.

Metamorphic rocks
■ Quartzite
■ Marble

Marble can be green, black, white, pink or a mixture of colours. It is a very beautiful rock and is used in the making of sculptures, headstones, fireplaces and ornaments. Connemara marble is green with black or white markings.

Marble
Limestone is changed to **marble** when it is 'cooked'.

Think of how a lump of dough can change to a lovely rock bun when it is baked in the oven!

Mt Errigal in Co. Donegal

Quartzite

Sandstone is changed to a metamorphic rock called **quartzite** when it is put under great pressure and squeezed. Quartzite is found at Mt Errigal in Co. Donegal and at the Sugar Loaf in Wicklow.

Quartzite is a creamy yellowish rock. You can see tiny sparkly pieces like the crystals in sugar. It is a very hard rock and usually stands up as tall mountains on the landscape.

Quartzite is used to make sandpaper and watches.

Key points

- A **metamorphic** rock has been changed due to great heat or pressure.
- Limestone is changed to make **marble**.
- Sandstone is changed to make **quartzite**.

Mining Rocks – An Extractive Industry

Key words

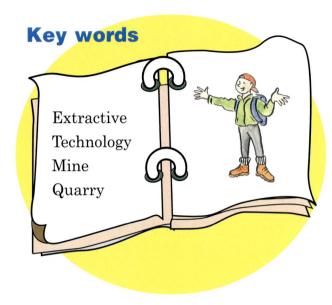

Extractive
Technology
Mine
Quarry

Rocks and their ores can be mined, quarried or drilled.

*A lead and zinc **mine***

A marble quarry

Once people decide that a particular rock can be of use to them they start to take it out of the ground. This is called an **extractive** industry. It brings money and jobs to the area. Today we have big machines (**technology**) that make it easier to extract the rock in large amounts.

Oil rig – drilling at sea

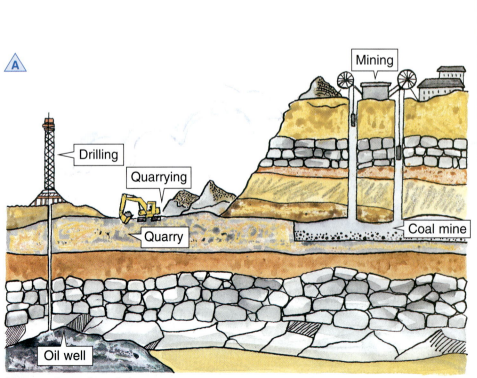

Extraction of rocks

Uses of rocks

Buildings

Coal fire

Plastic

Toothpaste

Clay butter dish

Cutlery

All kinds of tins

Jewellery

Paint

Look around you. Notice how many ways rocks are useful to you. Can you add to the list?

Key points

- Rocks are **extracted** from the ground.
- Some rocks are **mined**, others are **quarried** while others are drilled.
- The extraction of rocks creates jobs, bringing money to the local area.

REVISION EXERCISES

Write the answers in your copybook.

1   Rocks belong to one of three groups. In your copybook write down
    the groups.

    (a) I _____   (b) M _____   (c) S _____

2   Name two examples of igneous rocks.

3   In your copybook match each letter in column X with the number of
    its pair in column Y.

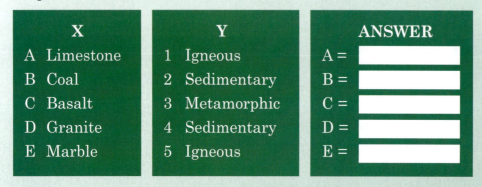

X	Y	ANSWER
A Limestone	1 Igneous	A =
B Coal	2 Sedimentary	B =
C Basalt	3 Metamorphic	C =
D Granite	4 Sedimentary	D =
E Marble	5 Igneous	E =

4   List **three** ways that rock, sand and gravel are used in Ireland.

5   Name any three minerals found in rocks.

6   Rocks are divided into **three** groups, sedimentary, igneous and
    metamorphic, depending on how they were formed. Choose **three**
    rocks – one from **each** group – and explain how each rock was formed.

7   Natural stone has many uses. Describe two of the uses of natural
    stone (rock) in your local environment.

8   List **five** examples of products that come from rocks and are often
    found in the kitchen.

9   Name two fuels that come from rocks.

10 Identify by looking at the map where the following rocks are found: granite, basalt, limestone, sandstone and marble.

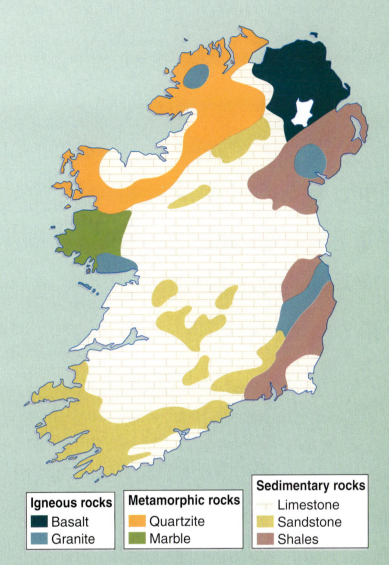

Igneous rocks
- Basalt
- Granite

Metamorphic rocks
- Quartzite
- Marble

Sedimentary rocks
- Limestone
- Sandstone
- Shales

11 Write out the paragraph below in your copybook and fill in the missing words from the list given.

> oil   basalt   building   soil   conditioner
> gas   Antrim   formed   igneous

The rocks of the earth are divided into three groups based on how they were _____. The liquid rock from inside the earth cools to form _____ rocks. The oceans of the world are made up of _____ rock. This is a hard rock. It is found in Co. _____. Rocks have many uses including _____ and _____ _____. Sometimes you get fuels like _____ and _____ from rocks.

Sea erodes rocks

Erosion changes the landscape

The world is made up of many different types of landscapes. No two places are exactly alike. Geographers like to notice and explain how these differences come about.

We have seen how plate movements create landscapes (Chapter 1). Now let's look at the forces that work on the surface of the earth to shape these landscapes. These are the forces of **weathering** and **erosion**.

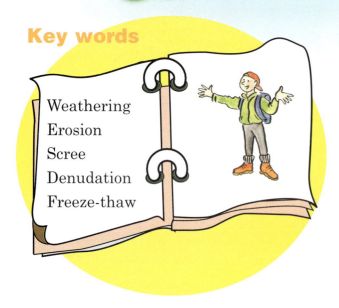

Weathering breaks down rocks

Forces of weathering

The breakdown of rock by the weather is called **weathering**. The weather – rain, heat and frost – helps to break-up rocks into smaller pieces called **scree**. The weather can also dissolve rock.

Forces of erosion

There are other forces that work on the surface of the earth. These forces include the work of rivers, ice, the sea, wind and human activity. These are called forces of **erosion**.

Denudation

Weathering and erosion work together, to break-up rocks and shape the landscape. Together they are known as forces of **denudation**.

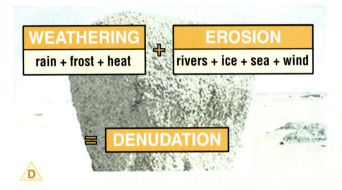

Types of weathering

Weathering breaks-up rocks in the following ways:

- Physical weathering – this means the shattering of rocks.
- Chemical weathering – this means the dissolving of minerals in rocks.

Physical Weathering

Let's look at an example of the physical breakdown of rocks by the weather. This is called frost shattering or **freeze-thaw** action.

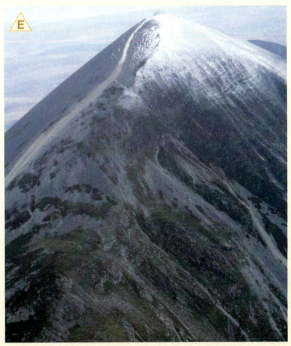

Frost action is common on mountains

Shattering of rock by freeze-thaw action

When water freezes and thaws in cracks in rocks, the rock may break-up. This is called frost shattering or **freeze-thaw** action. Let's take a look at what happens.

- Water seeps into a crack in a rock.

- When temperatures fall below freezing point, the water in the crack freezes. Ice takes up more space than water, so the ice forces the crack to widen.

- When this happens again and again, the crack widens till the rock splits and breaks away.

This type of rock breakdown happens on mountains where it is cold. This is where frost would be common.

Key points

- **Weathering** means the breakdown of rock.
- **Erosion** refers to the shaping of landscapes by ice, wind, rivers, the sea and human activity.
- **Denudation** is the wearing down of the landscape by weathering and erosion.
- Rocks are broken-up by **freeze-thaw** action.

Chemical Weathering

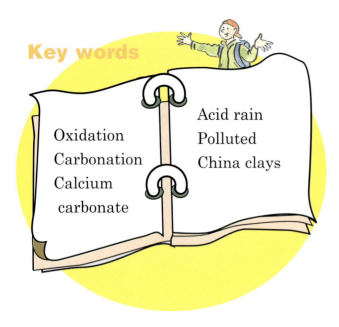

Key words

Oxidation
Carbonation
Calcium
 carbonate

Acid rain
Polluted
China clays

Chemical weathering happens when rainwater mixes with the minerals in some rocks. A chemical reaction takes place. The rocks are changed. This weakens them and they break-up. Let's look at two examples of chemical weathering: **oxidation** and **carbonation**.

Oxidation

If you left your bicycle out in the air and rain for any length of time you would soon notice how the metal would turn orange and eventually crumble. That is a good example of the weathering of metal by the process of **oxidation**. You would simply say the metal went rusty.

Oxidation happens in nature. Rocks with iron in them are weathered when rainwater and the oxygen in the air attack them. The iron crumbles and the rock breaks-up.

Carbonation

Carbonation happens when the gas, *carbon dioxide* (present in the air), mixes with rainwater and dissolves minerals in rocks. One mineral that is easily dissolved is **calcium carbonate**.

Calcium carbonate dissolves in rainwater in the same way a soluble tablet dissolves in water. The rock, limestone, contains calcium carbonate and so it is easily dissolved away.

You will learn more about the weathering of limestone in the case study on the Burren in Co. Clare.

Metal rusts in rainy weather

Humans and Weathering

Acid rain

Weathering is a natural process, but human activity can speed up the rate at which it happens. This is most obvious in cities where **acid rain** has caused the surface of some buildings to crumble and fall off.

Acid rain is caused when the air is **polluted** with the gases from car exhausts and chimneys. Gases from the burning of coal and oil are mixed with rainwater to make acid rain.

The Acropolis in Greece has been weathered more in the last 100 years as a result of acid rain than in all the years since it was built

The benefits of weathering for people

The weathering of rocks can have benefits (good results) for people.

Weathering helps to make soil. We can grow crops in soil and so we get food

China clays *are used to make clay pots and dishes. This gives work to people in the local area*

Key points

- Chemical weathering happens when rainwater reacts with minerals in rocks.
- **Carbonation** and **oxidation** are examples of chemical weathering.
- Human activity, resulting in **acid rain**, causes the weathering of rocks.
- Weathering of rocks can have benefits for people.

The Burren – The Weathering of Limestone

Key words

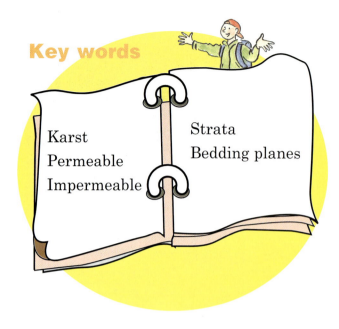

Karst
Permeable
Impermeable

Strata
Bedding planes

The Burren

The Burren area in the north of Co. Clare is an area with a unique landscape. It has a bare rocky surface that some people have compared to the surface of the moon. Underground you will find amazing caves and tunnels. This limestone landscape is known as a **karst** landscape. This landscape has come about because of the way rainwater has worked on the limestone rock in this area.

The Burren

There is very little soil covering the limestone rock in the Burren. Rainwater falling on the limestone can attack it directly. When the rain falls on limestone it dissolves it. This opens up cracks. The rainwater seeps downwards.

This is why you will find few rivers running along the surface in a limestone area. Water tends to disappear down through these cracks in the limestone. The rock is said to be **permeable**. This means it allows water to seep downwards.

> **Note:** Where water runs along the surface we say the rock is **impermeable**. It does not let water seep downwards.

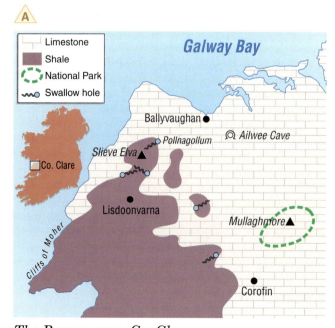

The Burren area, Co. Clare

Let's take a closer look at why limestone is so easily weathered by the rain.

- Limestone has joints or cracks in it. Rainwater can pass down through these cracks. It dissolves the rock as it goes down through it.

- Limestone is made from the remains of sea creatures. It contains their bones. These bones add calcium carbonate to limestone (your bones contain calcium carbonate too). This mineral dissolves easily in rainwater.

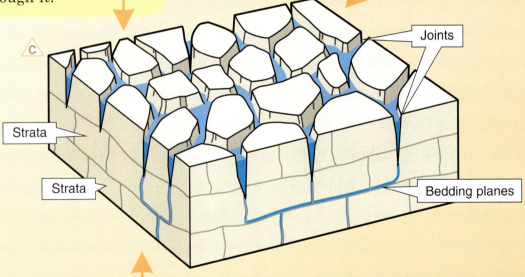

Joints

Strata

Strata

Bedding planes

- Limestone is a layered rock. A layer is called a **strata**. Rainwater can seep along the spaces between the layers. These spaces are called **bedding planes**. As it soaks along between the layers, it dissolves the limestone.

Key points

- The Burren is a **karst** landscape.
- Limestone is a **permeable** rock.

The Burren – The Results of Chemical Weathering

Key words

Swallow holes
Limestone pavements
Clint

Grike
Stalactites
Stalagmites
Karst landscape

Limestone pavements

Picture a pavement as a series of slabs of rock set alongside each other. Cracks or joints separate the slabs of rock. Each slab is called a **clint** and the cracks are called **grikes**.

Weathering by the process of carbonation is happening along the joints where the rock is weakest. The joints are widened to form grikes. Plants, such as gentian, find a cosy home in the sheltered spaces in the grikes. It is warm and dry here and plants more usually found in warmer places, such as the south of France, survive on the Burren.

Let's look in more detail at the features that result from the weathering of limestone. First we will look at features on the surface. These include:

- **Swallow holes**
- **Limestone pavements.**

Swallow hole

Swallow hole

A **swallow hole** is a hole into which a river disappears. Rainwater opens up the joint or crack in the limestone so that a stream is swallowed up. The stream running along the surface will now disappear down the widened joint (**grike**) and continue its journey underground. You will find a swallow hole at the place where a river crosses from impermeable rock to the permeable limestone rock.

Let's look at underground features. These include:

- Underground caves
- **Stalactites** and **stalagmites**.

Underground caves

As the river runs along underground it eats out tunnels and caves. The processes at work here are carbonation and river erosion. The Ailwee Caves are formed in this way.

The largest underground limestone cave system is in Mammoth Caves in Kentucky, USA.

Stalactites and stalagmites

Stalactites and **stalagmites** are found in underground caves.

Picture the rainwater working its way down through the joints in the rock, dissolving the limestone as it passes along. When a drop of water carrying the dissolved limestone gets to the roof of the underground cave, it hangs there. The water will dry off leaving a hard deposit (called calcite) behind. It looks like a straw. This is the beginning of a **stalactite**. If the drop splashes to the floor it will begin to form a **stalagmite**.

After millions of years of deposits building up, the stalactite will extend from the roof of the cave to join up with the stalagmite from the floor of the cave. In this case you get a pillar.

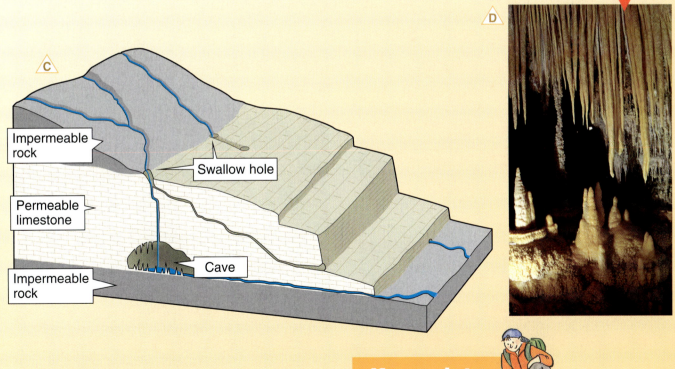

C

Impermeable rock

Swallow hole

Permeable limestone

Cave

Impermeable rock

D

Key points

- A **karst landscape** has both surface and underground features.

The Burren – Tourism

Key words

Tourism
Conflict of
interest

A tourist attraction

Many visitors visit the Burren area of Co. Clare every year. It has become an important area for **tourism**.

A conflict of interest

Tourism could be described as a mixed blessing. Some people are very happy to welcome visitors to the area. Others are not so sure it is a good thing. This is an example of a **conflict of interest**.

Castle in the Burren

Dolmen in the Burren

Honeysuckle nestling in a grike

Should the natural landscape of striking rocks and rare flowers be left undisturbed by busloads of tourists? Should the limestone be dug out and the space used for building visitor centres and car parks? What do the locals think? Are jobs more important than looking after rare plants and scenery?

Look at the picture that shows you the various interest groups discussing what should happen in the Burren.

How can everybody be satisfied? Is this possible? Perhaps some of the following could be done:

- Put a limit on the number of tourists visiting the area.
- Keep some areas totally natural and free of tourists.
- Build tourists centres and toilet facilities in towns which are not in the heart of the Burren e.g. Ballyvaughan. This won't spoil the natural beauty of the place or cause pollution.

Key points

- The Burren is an area of great natural beauty.
- The Burren has many **tourist** attractions.
- A **conflict of interest** arises when people disagree about how best to use an area of natural beauty.

REVISION EXERCISES

Write the answers in your copybook.

1 Explain the following terms: denudation, weathering and erosion.

2 Describe **two** ways in which you would know that a building in your area has been weathered.

3 Explain why freeze-thaw action usually happens on mountains.

4 Match each letter in column X with the number of its pair in column Y.

X	Y	ANSWER
A Chemical weathering	1 Flower of the Burren	A =
B Stalactite	2 Underground	B =
C Grike	3 Exploring limestone caves	C =
D Gentian	4 The Burren	D =
E Potholers	5 Surface	E =

5 Look at the diagram of underground limestone features. Which row shows the features shown with the correct letters?

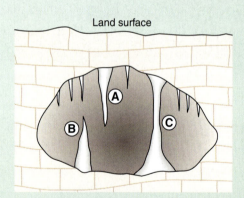

Land surface

(i)   A: clint, B: stalactite, C: stalagmite
(ii) A: stalactite, B: stalagmite, C: clint
(iii) A: stalagmite, B: stalactite, C: pillar
(iv) A: stalactite, B: stalagmite, C: pillar

6 In your copybook write the correct word for each of the statements below:
(a) The chemical weathering of limestone leads to cracks on the surface called **clints/grikes**.
(b) Underground features which hang from the roof of the cave are called **stalactites/stalagmites**.
(c) These features are common at the **Giant's Causeway/the Burren**.

7 Write out the paragraph below in your copybook and fill in the missing words from the list given.

> **erosion   scree   weathering   grikes**
> **denudation   minerals   Clare**

The laying bare of the earth's surface is called_____.
This is carried out by _____ and _____.
Rocks can be broken up into smaller pieces called _____.
When water reacts with _____ in some rocks a chemical reaction
takes place. This leads to the break-up of this kind of rock. An area
that has been weathered in this way is the Burren in Co. _____.
Here the joints of the limestone rocks have been widened to form

_____.

8 The most visited caves in Ireland are the Ailwee Caves, Co. Clare.
Write a postcard describing a visit.

9 Name **two** surface features found in a karst area. For **each** example
draw a diagram and explain how it was formed.

10 Copy the diagram given in question 5 in to your copybook. In the
case of **two** of the underground features explain how they formed.

11 Many tourists visit the Burren because of the attractions it offers.
List **three** different reasons why a visitor might go there. Then, in
the case of each reason given, explain its attraction.

12 Planning permission for a 'holiday town' in the Burren has been
requested by a local builder. He plans to build 60 holiday homes, a
leisure centre and a nightclub. The site is six miles outside
Lisdoonvarna in a quiet area.

Give two reasons
- Why local people might agree with the project.
- Why local people might disagree with the project.

13 Name **one** good effect and **one** bad effect of weathering.

4 Mass Movements

Key words

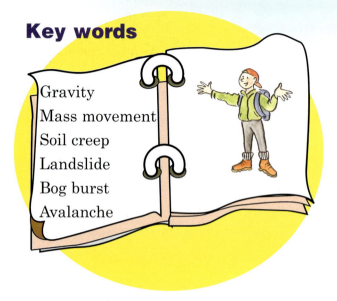

Gravity
Mass movement
Soil creep
Landslide
Bog burst
Avalanche

Rock and soil on hillsides move down the slope because of the pull of **gravity**. This type of movement is called **mass movement**. The term mass movement means that large amounts of material move downslope.

To understand the idea of gravity just picture what happens if you park a car on a slope and forget to apply the handbrake. The car will roll down the hill. Gravity is the force that causes this movement.

Let's look at four types of mass movements: **soil creep**, **landslide**, **bog burst** and **avalanche**.

Soil Creep

Soil creep is a slow mass movement. You would not notice this happening because it is such a small movement. Grain by grain, the soil and pieces of rock creep down the hill, pulled along by gravity. You know it has happened because after a long time fences on slopes begin to lean over. Look at diagram **B** of a slope to see other effects of soil creep. This type of movement is common on very gentle slopes.

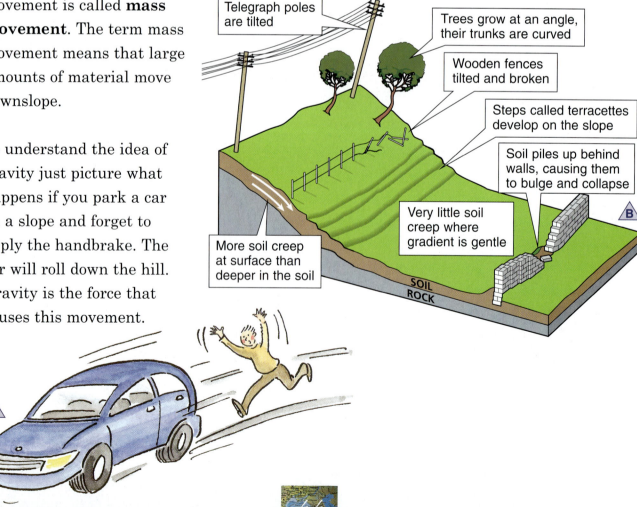

Telegraph poles are tilted

Trees grow at an angle, their trunks are curved

Wooden fences tilted and broken

Steps called terracettes develop on the slope

Soil piles up behind walls, causing them to bulge and collapse

Very little soil creep where gradient is gentle

More soil creep at surface than deeper in the soil

SOIL
ROCK

B

A

Landslide

A very fast movement of land is called a **landslide**. During a landslide, rock and soil slide down the slope bringing houses and even roads with it. Landslides are common on steep slopes. They happen suddenly and without warning. They are described as dry movements because rain is not always a trigger in this type of movement.

What causes a landslide?

A landslide can be triggered (set off) by natural causes such as:

- An earthquake, for example, in San Francisco in 1989.
- The waves which constantly erode into the bottom of a cliff leaving the top of that cliff to collapse and slide. This has happened along the coastline of Co. Antrim.

A landslide can also be triggered by people when they:

- Cut into a slope for building houses e.g. the shanty towns of India and Brazil.
- Cut into a slope to build roads e.g. Co. Antrim.

Damage caused by landslides – catastrophe in India

A shanty town collapsed killing scores of residents in the overcrowded city of Mumbai in India. Here, many people live on slopes around large cities. Their houses are unplanned and very crowded. They are often made of flimsy materials such as corrugated iron, cardboard and old wood. For years the city authorities had given out leaflets warning of possible landslides but no attempt was made to move the people.

More than 5 million people live in the slums of Mumbai. They are at risk from diseases, flooding and landslides.

Key points

- **Mass movement** is the movement of large amounts of loose rock and soil.
- The pull of **gravity** causes mass movement.
- **Soil creep** is a very slow mass movement.
- **Landslides** are very fast mass movements.

Bog Bursts

Key words

Bog burst
Avalanche
Vegetation
Terraces

A third type of mass movement is called a **bog burst**. It is described as a wet movement of bog or peat down a slope.

This can be a fast or slow movement. It depends on the steepness of the slope and the amount of water in the peat. A section of a bog can become saturated with water. It is now heavier and so it slides down the slope very suddenly and quickly. Bog bursts are common in the West of Ireland.

Heavy rain triggers bog burst in Co. Mayo

Peat Moves

Emergency services were working at full stretch yesterday to clear the area outside Belmullet. The mass movement, which happened after hours of torrential rain, blocked a road, encircled homes and damaged two bridges cutting off a village at one point. Thousands of tonnes of material choked up the roadway, cutting off up to 25 houses close to the village of Pollathomas. Luckily no one was injured. 'It has certainly proved that peat can move' one bystander said. It showed how unstable peat can be. Others commented on the fact that over 300,000 cubic metres of peat had been removed for the Knock-Claremorris bypass and nobody seemed to know what was done with it.

From *The Irish Times*

Avalanche

An **avalanche** is a mass movement of snow. Avalanches happen on slopes greater than 25°. They happen very quickly and can be triggered by a sudden noise (e.g. an explosion or an earthquake) or after heavy snowfall.

To protect themselves on some slopes, skiers now use the ABS – avalanche balloon system – which is like an inflatable lifejacket. It is red and easily seen in snow. It has a cord which, when pulled, inflates a balloon that then brings the wearer to the surface where they can be seen.

Mass movements are more common where:

- There is heavy rainfall, particularly after a dry period.
- Slopes are steep.
- There is little **vegetation** (plants and trees) on the slope.
- Earthquakes are common.
- People have cut into a slope for roads or railway lines.

How to respond to mass movements

- Avoid building houses on steep slopes.
- Plant trees to bind the soil together.
- Cut **terraces** (steps) into the side of the mountain – as in China – this reduces the movement of the soil.
- Place barricades or fences on mountain slopes.

Key points

- A **bog burst** is the mass movement of peat.
- An **avalanche** is the mass movement of snow.
- Mass movements can be set off by nature or by people.
- Human effort can reduce mass movements.

REVISION EXERCISES

Write the answers in your copybook.

1   Which of the following is shown in this diagram?
   ● A volcanic eruption
   ● Acid rain
   ● A landslide
   ● Soil creep

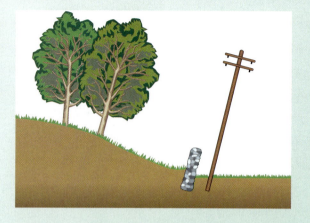

2   This diagram shows an example of:
   ● Bog burst
   ● Soil creep
   ● Landslide
   ● Avalanche

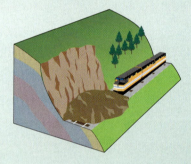

3   Which one of the following is an example of mass movement?
   ● Traffic going into the city
   ● Soil creep
   ● Rainfall
   ● Vegetation

4   Sometimes, soil or loose rock can move very quickly down a slope.
   This is called:
   ● Soil creep
   ● Solar creep
   ● Landslide
   ● Bog burst

5   Write the **incorrect** answer for each of the following in your copybook.
   (a) Mass movements are influenced by **gravity/light.**
   (b) Soil creep is an example of a **slow/fast** mass movement.
   (c) Avalanches occur on **steep/gentle** slopes.

6   Write out the following paragraph in your copybook. Use the list given in the box to fill in the missing words.

landslide rock wet slow gravity faster soil

Mass movements are the movements of large amounts of _____ and _____ material downslope. They are controlled by the force of _____. They can be fast or _____. A bog burst is a _____ movement. A _____ is an example of a fast mass movement. The steeper the slope, the _____ the movement.

7   Read the newspaper heading and answer the questions below.
Over 50 Killed as Landslide Hits Shanty Town in India
(a) What is a landslide? Explain how it happens.
(b) Explain briefly why so many people would be killed by a landslide in a poor country.

8   In your copybook match each letter in column X with the number of its pair in column Y.

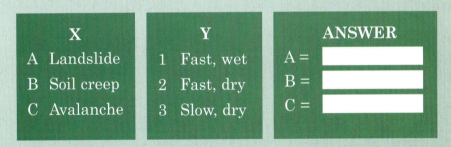

X	Y	ANSWER
A Landslide	1 Fast, wet	A =
B Soil creep	2 Fast, dry	B =
C Avalanche	3 Slow, dry	C =

9   Draw a clearly labelled diagram showing the effects of soil creep on a slope.

10   Describe **two** ways of reducing the damage that a mass movement may cause on a hillside.

11   Explain why a landslide that happens in a poor country may have more serious effects than one that occurs in a rich country.

12   Avalanches only happen under certain conditions. Name **two** of these conditions.

13   Name **two** differences between soil creep and a landslide.

5 Rivers

Key words

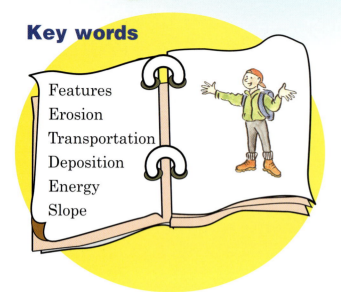

Features
Erosion
Transportation
Deposition
Energy
Slope

Shaping the Landscape

A river shapes **features** on the landscape just like a sculptor shapes features in a block of stone.

A river shapes the landscape by the work of **erosion**, **transportation** and **deposition**. To do this work the river has to have **energy**.

Erosion

The work of **erosion** involves the wearing down and removal of soil and rock from the bed and banks of the river.

Transportation

The work of **transportation** means moving the soil and rock to a part of the river further downstream.

Deposition

The work of **deposition** means dropping the soil and rock someplace along the course of the stream.

Energy

People need **energy** to work. Rivers also need energy. Anything that gives the river energy will allow it to do its work of erosion and transportation. If the energy of the river is lessened it will not be able to carry its load. It will drop it. People would do the same. When you run out of energy you put down your load!

A **slope** will give energy to a river. A river will be able to run faster down a steep slope. Isn't this true of you? Picture yourself on your bike! Won't you travel faster down a steep slope? When the slope evens out, you will slow down.

Remember
- A fast flowing stream has energy to erode and transport its load.
- A slow flowing stream has less energy and will deposit its load.

Adding more water to a stream will also give energy. When a river becomes deeper it can run faster. You might have noticed your local stream running much faster when it is in flood.

The different stages of a river.

On its journey to the sea, we say that a river passes through three stages. The first stage is **youth**, then **maturity** and then **old age**, just like a person. You find different features at each stage.

Before looking in detail at the features made by a river, it is important to get to know some of the basic terms and ideas which you will come across when studying this topic.

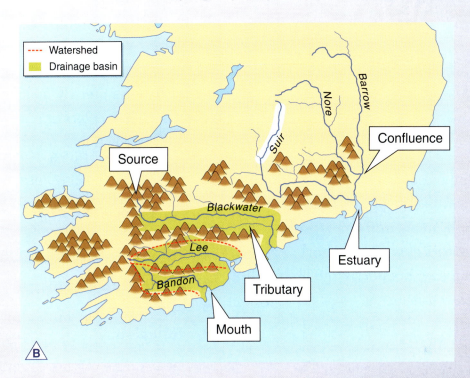

More key words!

- The **source** is the place where a river begins. The **source** is usually in the mountains.
- Other smaller streams called **tributaries** join the river as it makes its journey to the sea.
- The **mouth** of a river is where the river enters the sea.
- The **estuary** is the part of the river that mixes with the sea and is tidal (it rises and falls with the tide).
- Rivers **drain** the land. The rain falling in an area finds its way to the river to be drained away into the sea.
- The whole area drained by a river is called a **drainage basin**.
- Geographers name the stages of a river's journey as **youth**, **maturity** and **old age**.

Key points

- The work of rivers is to **erode**, **transport** and **deposit** material such as rocks, sand and silt.
- Rivers have three stages on their journey to the sea: youth, maturity and old age.

First Stage – Youth

Key words

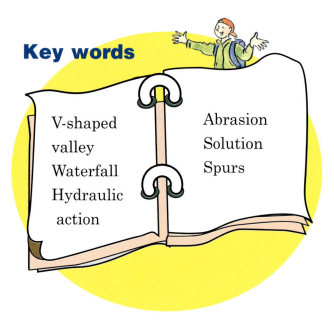

V-shaped valley
Waterfall
Hydraulic action

Abrasion
Solution
Spurs

Erosion

A river starting out on its journey is said to be in the youthful stage of life. Like a child, it is small, lively and has lots of energy. It uses this energy to remove anything in its way. Erosion is the main work of the river during the youthful stage. The river erodes in different ways. These ways are called 'the processes' of erosion.

Features in the youthful stage

In its youthful stage, high up in the mountains, the main work of a river is erosion. The river forms:

- A **V-shaped valley**
- **Waterfalls**.

V-shaped valley

At the beginning of its course the river is small. It flows quickly as it comes down the mountain. It has lots of energy, which it uses to cut down into its bed. In this way the bed is deepened. The sides of the valley rise up steeply from the river forming a narrow **V-shaped valley**.

This V-shaped valley is formed by:

- **Hydraulic action**
- **Abrasion**
- **Solution**.

 The processes by which the river erodes its bed and banks

Hydraulic action: This is the force of the water. It is the same process as a car wash! The force of the water loosens rocks along the bed and banks of the river.

Abrasion: The word abrasion means to scrape. The river uses the stones it is carrying to scrape away the bed and banks of its channel.

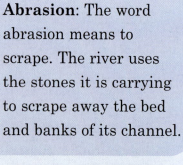

Solution: The water dissolves some minerals in the underlying rocks and washes them away.

Because it is small the river is unable to remove spurs that lie across its path. Spurs are lumps of hard rock that stick out across the valley. The river wanders around these spurs. Look at the picture to see the river as it winds around these interlocking spurs.

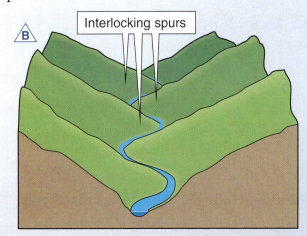

Waterfall

A **waterfall** is a steep section along the course of the stream. The river falls over the steep part forming a waterfall. It is a feature of erosion.

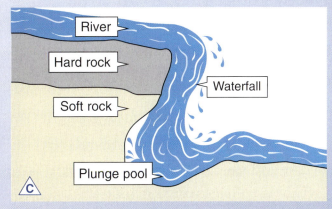

Soft rock is worn down. Water tumbles over the drop

The steep part happens because a band of hard rock runs *across* the bed of the stream. The river runs over the hard rock and onto the softer rock.

The river erodes the area of soft rock and wears it down to a lower level. The river tumbles down onto this section. A waterfall is formed.

Example: Glencar Waterfall, Co. Leitrim

Glencar Waterfall in Co. Leitrim

Key points

- In its youthful stage, the main work of the river is erosion.
- The river erodes by **hydraulic action**, **abrasion** and **solution**.
- A **V-shaped valley** and a **waterfall** are formed by river erosion.

Middle Stage – Maturity

Key words

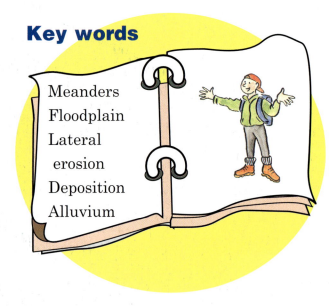

Meanders
Floodplain
Lateral erosion
Deposition
Alluvium

Transportation

Now that the river has eroded rocks and pebbles, it transports its load downhill.

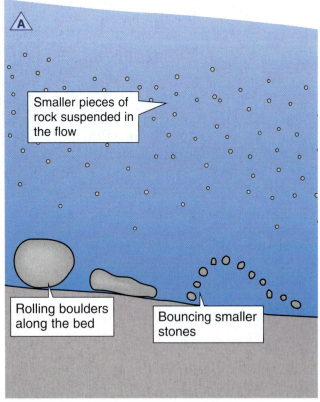

Smaller pieces of rock suspended in the flow

Rolling boulders along the bed

Bouncing smaller stones

The way a river transports its load depends on the size of the rock pieces

Maturity

The river now moves into the middle section of its course. We call this the mature stage.

The river is bigger now because tributaries have joined it. At this point the slope is levelling out a bit. Two features found at this stage are:

- **Meanders**
- A **floodplain**

Meanders

In the middle or mature stage of a river's journey to the sea it will flow in broad loops called **meanders**. Look at picture C. You find meanders on the River Shannon.

As the river flows around a bend in these loops it will flow fast on the outside of the bend. It will slow down on the inside. Think of yourself cycling around a bend. You would pedal faster on the outside. You would brake and slow down going around the inside of the bend.

Erosion takes place on the outside bend where the flow of the river is fast and the river has lots of energy. This is called **lateral erosion**.

Deposition takes place on the inside of the bend.

Flood plain

A floodplain occurs in the mature stage of a river's course

Meanders occur in the mature stage of the river's course

A floodplain

A **floodplain** is an area of level land alongside a stream. It is built up when a river overflows, spreads out and drops its load of sediment. You find a floodplain in the middle and final stages of a river's course.

As a stream begins to slow down it will drop its load on its bed. It now runs on top of these deposits. It can flood more easily now because it is nearer to the top of its banks. After heavy rain the level of the river will rise further. The river may flood and spread out over its valley. It will drop a fine soil called **alluvium** on the floodplain. Alluvium makes a very fertile soil.

Key points

- A river forms **meanders** and **floodplains** in its middle (mature) and final (old age) stages.

Last Stage – Old Age

Key words

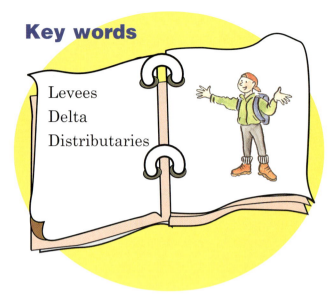

Levees
Delta
Distributaries

In the old age stage, the river is flowing slowly because the slope has levelled out. The river is carrying a large amount of mud and sand which it eroded along its course. The river will meander broadly. It will have a floodplain. Sometimes the river will have built up **levees** along the banks.

Levees – a feature formed by deposition

Levees are high banks that run along the side of the river. They are made when a river overflows. The river drops its heaviest load when it first overflows. Sand and gravel will pile up to form a ridge all along the banks of the river.

The Mississippi River and the River Shannon have natural levees. The Ganges has some artificial or man-made ones. In this case, the people have built up the banks of the river with concrete or boulders so that the river is held back when the level of the water rises.

Floodplains, **levees** and meanders can occur in the middle or old age stages of a river's course.

A

Raised banks (levees)

Levees occur in the old age stage of a river's course

Delta – a feature of deposition

A **delta** is shaped like a triangle. It is found at the mouth of the river. It is made of very fine deposits of rock called silt and is crossed by many streams.

A delta is formed when a river finally empties into the sea and deposits its load. If the tides or sea currents are not strong enough to sweep the load away, the deposits will build up at the river's mouth. They will block the flow of the river. The river has to find a way through these deposits so it will split into separate channels called **distributaries**.

Examples of rivers that have large deltas are the Mississippi in the USA, the Ganges in Bangladesh and the Nile in Egypt.

This satellite picture shows the Nile Delta and its distributaries flowing into the Mediterranean sea

Summary

Stage of youth	Fast flowing stream in a narrow V-shaped valley	Features: waterfalls and V-shaped valleys (erosion)		*V-shaped valley*
Stage of maturity	Meandering stream in a widening V-shaped valley	Features: meanders (lateral erosion) and floodplains (deposition)		*Meandering stream*
Stage of old age	Slow flowing stream in an almost flattened V-shaped valley	Features: floodplains, levees and deltas (deposition)		*Wide floodplain*

Key points

- The work of a river in old age is mainly to deposit its load.
- Two features of deposition found in the old age stage are **levees** and **deltas**.

Human Activity and Rivers

Key words

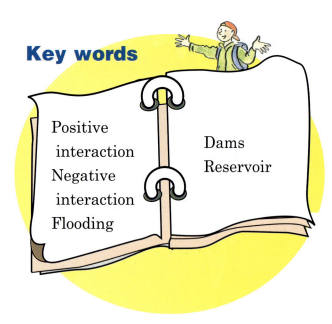

Positive interaction
Negative interaction
Flooding

Dams
Reservoir

 Positive interaction

- Farmers grow crops on fertile soil irrigated by water from rivers.

People deal with or interact with rivers in **positive** and **negative** ways.

- People use rivers for fishing.

- Hydroelectric power stations get their energy from rivers.

- People use rivers for leisure activities.

 Negative interaction

- Throwing waste into rivers from homes can destroy fish and plant life.
- Pouring hot water from factories into rivers can kill many fish.
- Allowing slurry from farms to seep into rivers can destroy the fish life in them. Look at picture **E**.

Focus on flooding

Flooding has been happening more often in recent years. It may be because there is more rainfall (due to global warming which you will learn about later). But it may also happen because houses, roads and factories are being built on floodplains.

Building settlements on floodplains can be risky for two reasons:

1   A floodplain is a natural feature. A river will overflow from time to time because that is its nature!
2   By putting a concrete surface on a floodplain, rainwater does not seep gradually into the ground. It runs into drains and feeds into local rivers. The levels of these rivers can rise very quickly and they can overflow flooding towns.

The flooding of a built up area happened in 2003 at Leixlip in Co. Kildare when the River Liffey overflowed. Parts of Dunboyne in Co. Meath were flooded when the River Tolka overflowed.

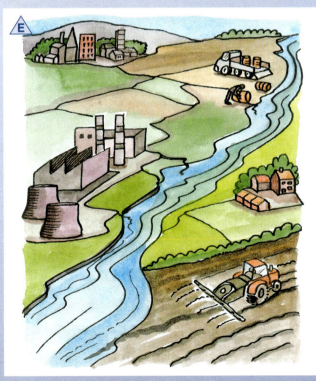

Notice negative (bad) uses of rivers

Flooding causes great damage. There may be loss of life. In the countryside farmland can be ruined. In built up areas houses can be swept away in floodwater. Furniture and carpets can be ruined. The cost of repairs can be huge.

Managing the flow of a river

You have learned about the danger of flooding to people living on a floodplain. Can anything be done to manage the flow of rivers?

The following steps can be taken:

● Build up the banks of rivers (artificial levees) to hold the water in when the level of the water rises.

● Build **dams** to hold water back in a **reservoir** (artificial lake). The water can be let out when the danger of the flood lessens.

● Take more care about building housing estates and roads on floodplains. Without a covering of concrete, the rainwater would seep into the ground more gradually and drain more slowly into the rivers.

Key points

● Rivers are very useful to people.
● People have abused rivers by polluting them.
● Rivers in flood can threaten life and property so efforts could be made to manage this flow.

REVISION EXERCISES

Write the answers in your copybook.

1 In your copybook match each letter in column X with the number of its pair in column Y.

X	Y	ANSWER
A Source	1 Starting point of a river	A =
B Mouth	2 Youthful stage	B =
C Dam	3 Where the river ends	C =
D Distributary	4 River breaks up into smaller streams	D =
E Waterfall	5 Prevents flooding	E =

2 Explain the following terms:
  ● Hydraulic action
  ● Abrasion
  ● River's load

3 Describe with the help of a diagram **one** feature formed by river erosion **or** one feature formed by river deposition.

4 Copy the diagram of a meander into your copybook. Mark the places where erosion is happening. Mark the places where deposition is happening.

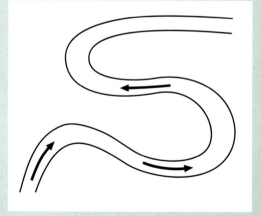

5 Explain with the help of a diagram **two** ways that a river transports its load.

6 Rivers have always been useful to humans. Describe **three** ways in which people use rivers.

7 Flooding frequently occurs along the course of a river. Explain **one** way in which flooding damages an area and **one** way that flooding can benefit an area.

8   Describe one way that people have managed the flow of a river.

9   Name and describe **one** way rivers cause problems for people.

10 Use the maps of Ireland provided at the end of the book to make the link between the rivers and the towns in the chart opposite.

River	Town
Boyne	Drogheda
_ _ _ _ _ _ _ _	Galway
_ _ _ _ _ _ _ _	Dublin
Lagan	_ _ _ _ _ _ _ _
Lee	_ _ _ _ _ _ _ _
_ _ _ _ _ _ _ _	Enniskillen
Suir	_ _ _ _ _ _ _ _

11 Complete the diagram. Write your answers in your copybook.

Stage	Valley shape	Features
	Fast flowing stream in a narrow V-shaped valley	Features: waterfalls and V-shaped valleys
Stage of Maturity		Features: meanders (lateral erosion) and floodplains (deposition)
Stage of Old Age	Slow flowing stream in an almost flattened V-shaped valley	

12 Write out the paragraph below in your copybook and fill in the missing words from the list given.

> **alluvium   youthful   hydroelectricity**
> **mountains   waterfall   tributaries**

Rivers usually begin in _____. They flow downhill and are joined by smaller streams called _____. At this stage they are fast flowing and are said to be in the _____ stage. If they flow over areas of hard and soft rock they erode the soft rock forming a _____. Rivers are very useful to humans. Factories get _____. This allows them to run machines. Farmers use the very fertile _____ soil to get high amounts of crops.

6 The Sea

Key words

Cliffs
Caves
Erosion
Compression
of air

Bay
Headland
Arch
Stack

The Landscape of Coastlines

You may have played on broad sandy beaches, looked over steep **cliffs** or explored sea **caves**. These features and others are formed when waves erode and deposit rock and sand along the shore.

Let's look first at the features of erosion.

Features of erosion

Erosion means the wearing away of the land. The sea is very good at this, especially during violent storms when the waves have loads of energy for erosion. The wind whips up the waves and throws them against the **cliffs**. Under this type of attack, the rock breaks off and is swept away by the sea.

The amount of erosion depends on the strength of the waves and the softness of the rock. Soft rock will be eroded more quickly than harder rock.

You will remember from your study of rivers that moving water erodes by the force of:

- *Hydraulic action*: The force of water.
- *Abrasion*: The waves using rocks to wear down other rocks.
- *Solution*: Water dissolving rocks.

The sea erodes in exactly the same way. It also erodes the rock by a process called **compression of air**. When waves rush into a cave, the air in the cave is squeezed or compressed. When the wave is sucked back into the sea, the air expands again in the cave. This happens over and over again. The rocks break up under this pressure.

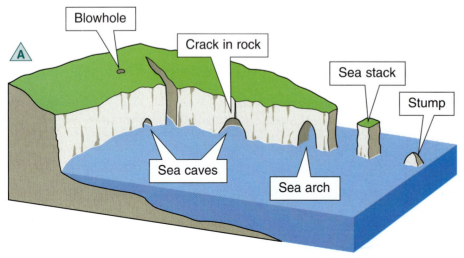

A

Blowhole

Crack in rock

Sea stack

Stump

Sea caves

Sea arch

Features formed by sea erosion

Waves attack the cliff. This forms a notch

The notch gets deeper as erosion continues

The overhanging section falls down

Cliffs

A **cliff** is a steep section of the coast. It is under attack by the waves, especially during storms. The waves attack the bottom of land where it slopes down to the sea. Then like rivers, the waves attack using the force of hydraulic action, abrasion and solution.

Rocks are removed and a small notch is formed. Erosion continues until the section that hangs above the notch cannot be supported. It then collapses. The slope becomes steeper. This is a cliff.
Examples: Cliffs of Moher, Co. Clare; Slieve League, Co. Donegal

Caves

A **cave** is a feature of erosion. It is an opening in the cliff face. It is formed when the waves find a weak spot, such as a crack, and erode it.
Example: Portsalon, Co. Donegal

Arches and stacks

When two **caves** form back to back on a **headland** they may meet to form an **arch**.

Erosion continues as the sea moves through the arch. The spray will attack the roof of the arch. Eventually the roof will collapse and a sea **stack** will form. Example: Sceilig Mhicil, Co. Kerry

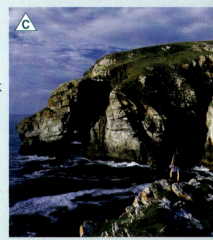

Sea arch

Bays and headlands

A **bay** usually sits between two **headlands**. They form on a coastline made up of different types of rock.

The incoming waves attack the weaker rock more easily making a bay. Headlands, the areas made up of stronger rock, jut out towards the sea on either side of the bay. Examples: Dublin Bay; Galway Bay

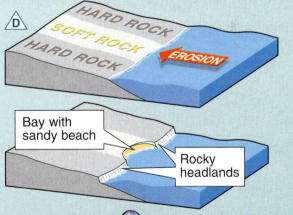

Bay and headland

Key points

- Waves erode the coastline especially during storms.
- **Cliffs**, **caves**, **arches** and **stacks** are formed by sea **erosion**.
- **Bays** and **headlands** are formed where you find rocks which differ in hardness.

Losing Land to the Sea

Key words

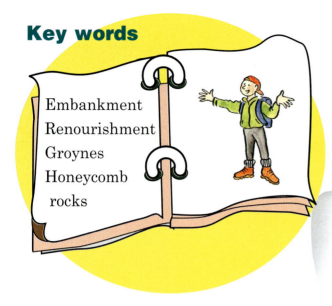

Embankment
Renourishment
Groynes
Honeycomb
 rocks

When land is lost to the sea during erosion the following can happen:

- Roads collapse
- Farmland is lost
- Houses fall
- Railway lines collapse.

The coast is constantly under attack. Strong waves, especially during storms, lash against the coast and remove some of it.

Here over 2 metres of land have been lost since 1999. Steps, built in 1999 by the council to help people onto the beach, are now in the middle of the sea. Roads have been swept away. An **embankment** to protect the coastline was built but even that is now in need of repair.

A

LOUTH

Louth – the smallest county is getting smaller

As long as it continues to lose an acre a year to the sea, Co. Louth, Ireland's smallest county, will get even smaller!

'Forty per cent of the coast in Louth is suffering a high degree of erosion,' according to a local engineer.

A stark example of the speed at which this erosion is taking place can be seen between Clogherhead and Blackrock.

A study of the 94km coastline has looked at areas in need of coastal protection. It has examined the role of easterly winds that batter the coastline during storms. It has highlighted the threat to the sand dunes which are in danger of being washed out to sea by the waves. A plan to protect the coast from attack by the sea could cost the government up to 5 million euros.

Defending the coast from attack

To protect the coast from attack, we must try to break the waves and so lessen their power.

One method is called a beach **renourishment** scheme. This involves bringing large amounts of sand to a beach. The idea is that the sand will act as a barrier and slow down the incoming waves. The coastline behind the barrier will then be protected.

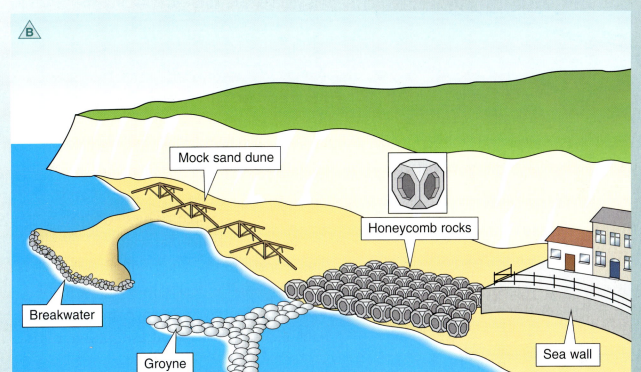

Mock sand dune

Honeycomb rocks

Breakwater

Groyne

Sea wall

Protecting the coastline

Another way to lessen the damage caused by violent waves is to build **groynes**. At Rosslare Strand, Co. Wexford groynes have been built. These 'walls' break the power of the wave and trap the sand and shingle. This method has been used to protect beaches along the coastline near Marbella in Spain.

Honeycomb rocks, which are artificial rocks, can be placed on the beach to stop the sea eroding the coastline.

Protecting the coastline is a very expensive exercise and needs funding from the government.

Key points

- The Irish coastline has been seriously eroded in places.
- Measures have been taken to protect Ireland's coastline.

Deposition by the Sea

Key words

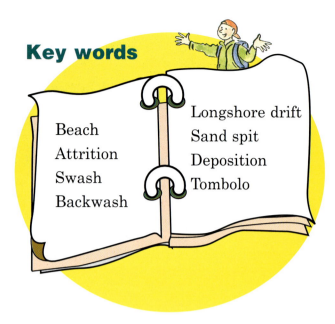

Beach
Attrition
Swash
Backwash

Longshore drift
Sand spit
Deposition
Tombolo

Rocks along the coast are eroded by the waves and swept out to sea. The broken off rock is moved about in the water. The rocks and stones are broken down further as they knock against each other. In this way, the rocks are made smaller and are rounded. Pick up a fistful of sand, shingle or stones from a **beach** and you will see this.

The rounding of rocks as they are knocked together is called **attrition.**

Swash and backwash – the transport system of the sea

The wind and the tides drive waves and their load of stones towards the shore. When the waves reach the shallow water of the shore they lose their power. They 'break' and run up the beach. This movement is called the **swash**. The swash moves up onto the shore at an angle driven by the wind.

The water flows back down the beach towards the sea again. This is called the **backwash**. The slope of the beach drives this action.

This movement of water up and down the slope of the beach is constant. Try standing on the edge of the water facing the sea. You will see the waves coming towards you and flooding around your feet. Wait a few moments and you will notice the water sucking back out to sea to be followed a few moments later by the next wave. The process is never ending.

The swash of water carries sand and shingle up the beach. The backwash carries some of this material back down the beach into the sea. Sand and shingle is moved along the beach in a zigzag motion as it is shifted by the swash and backwash. This movement is called **longshore drift**.

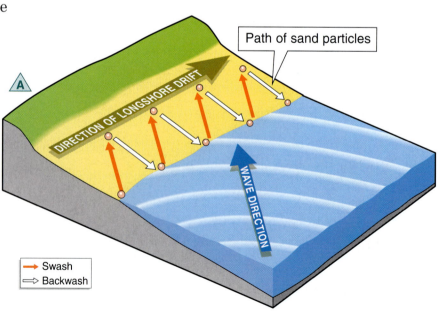

Features of deposition

Beach

A **beach** is an area of sand, pebbles or stones along the shore. It lies between the high-tide mark and the low-tide mark. It is formed by the swash and backwash movements of the waves.
Example: Curracloe, Co. Wexford

Sand spit

A **sand spit** is a ridge of sand or shingle that juts out into the sea. It is formed by **longshore drift** that carries the sand and shingle along the coast.
Example: Portmarnock, Co. Dublin

Sand bar

A sand bar is a **sand spit** that stretches across a bay to connect the two sides of the bay. The lake behind the sand bar is called a lagoon.
Example: Lady's Island in Co. Wexford

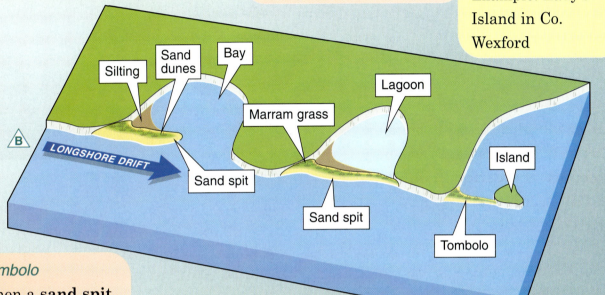

Tombolo

When a **sand spit** juts out into the sea and links up with an island, it is called a **tombolo**.
Example: Sutton, Co. Dublin

Sand dunes

Sand dunes are mounds of sand at the back of a wide beach. A sand dune is formed by the wind. The fine sand is whipped up from the beach and carried inland. The wind drops this fine sand when it loses energy. The sand builds up into sand dunes. Example: Rosses Point, Co. Sligo

Protecting beaches and sand dunes

Longshore drift moves beach material along the shore in a zigzag manner. In this way beach material is always being constantly moved along. To protect the coastline from the loss of beach or coastal material down along the shore people have:

- Built groynes (see diagram B page 65).
- Planted marram grass on the sand dunes.
- Banned people from taking sand from beaches.

Key points

- Waves transport sand and shingle onto a beach by the **swash** and **backwash**.
- **Deposition** takes place when the power of the waves is lessened.
- A **beach** is a feature of **deposition**.
- Bars, **spits** and **tombolos** are all ridges of **sand** and shingle deposited by the sea.

Humans and the Sea

Key words

Transport
Recreation
Fishing
Power
Fuel
Polluted

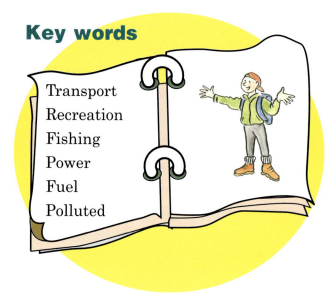

The sea is a great friend to humans. It offers us many advantages such as **transport**, **recreation**, **fishing**, **power** and **fuel**.

Power

Waves can be harnessed for power. This means their power can be used to make electricity for industry, homes and farms.

Fishing

Many coastal communities rely on the sea for their livelihoods. The sea provides us with fish, seaweed and salt.

Fuels

Oil and gas are found under the sea floor. People have built rigs and drilled for oil and gas under the sea floor.

Recreation

The seaside is a great place to relax. Sandy beaches and high waves used for surfing are delightful resources for tourist and locals alike.

Transport

Moving goods in boats across the oceans is one of the cheapest and easiest ways of transport. There are no roads to build and no traffic congestion most of the time. It is a vital link for trade between places.

People abuse the sea

The sea is a precious resource which has not always been treated with enough care. Human activities have damaged coastlines and **polluted** the sea.

Cleaning up after an oil spill

Swans feed from a raw sewage pipe at Broughty Ferry

A tanker spills oil off the coast of the Shetland Isles

Birds Killed!

BirdLife International today announced that it believes 10,000 to 15,000 birds have been killed by the Prestige oil spill. The spill happened along the Galician coastline in north-west Spain. They warned that the second wave of oil now heading ashore will be worse than the first. 'The Spanish population of guillemot has been hardest hit by the Prestige oil spill', a spokesperson said. 'We predict the guillemot is now very likely to become extinct as a breeding bird in Spain. If this happens the Prestige oil spill will be remembered as a tragedy for Spain's wildlife as well as for its people'.

Key points

- People have used and misused the sea.
- Good uses of the sea include **transport**, **recreation** and **fishing**.
- **Pollution** of the sea is a misuse of the sea.

REVISION EXERCISES

Write the answers in your copybook.

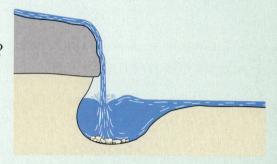

1 What is the feature shown in the diagram called?
   ● Beach   ● Esker
   ● Waterfall   ● Delta

2 Describe simply how one feature listed in question 1 is formed.

3 Which one of the following features is formed by deposition from the sea?
   ● Sand spit   ● Ox-bow lake
   ● V-shaped valley   ● Sea stack

4 Name **two** features of coastal erosion and **two** features of coastal deposition.

5 In your copybook match each letter in column X with the number of its pair in column Y. One pair is completed for you.

X	Y	ANSWER
A Cliff	1 Sea erosion	A = 1
B Stack	2 River erosion	B
C Beach	3 River deposition	C
D Waterfall	4 Sea erosion	D
E Delta	5 Sea deposition	E

6 Explain the terms:
   ● Hydraulic action   ● Abrasion
   ● Solution   ● Compression of air

7 In your copybook complete the statements:
   (a) The type of vegetation used to stop sand dunes moving inland is known as _____.
   (b) An area of sand and shingle found between high and low tides is called a _____.

8 Find the following places on the map of Ireland provided in this book:
   - Dublin Bay
   - Howth Head
   - Bray Head
   - Cliffs of Moher

9 The sea has many uses to us. Name **four** of those uses and write a note about each use (hint: transport, food, sport, energy, fuels). Use actual examples wherever possible.

10 Pollution is a major problem in our seas today. This pollution comes from many sources, including farms, factories and tourists. Explain how **each** adds to pollution of the sea.

11 Write out the paragraph below in your copybook and fill in the missing words from the list given. The words are in the box, but are not in the correct order!

> **waves   swash   sand   marram   shingle
> backwash   sand dunes   longshore drift   bigger**

The _____ approach a coastline. The stronger the wind the _____ the wave. The breaking wave rushes up the shore carrying _____ and _____. This is called the _____. Some of the material runs back into the sea as the _____. This is called _____ _____. Sometimes the sand is blown inland where it builds up to form _____ _____. Sometimes _____ grass is used to bind the sand and stop it from being blown away.

12 Describe with the help of a diagram **two** methods used in Ireland to stop the coast being eroded inland.

13 How do groynes prevent longshore drift?

Key words

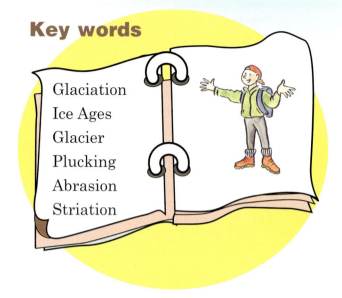

Glaciation
Ice Ages
Glacier
Plucking
Abrasion
Striation

Let's look at how ice shapes the landscape. This action is called **glaciation**.

Ice Ages

There were long periods in the past when Ireland was covered with snow and ice. The Irish landscape looked like that of Greenland today. These very cold periods were called **Ice Ages**. There were two Ice Ages in Ireland. The last Ice Age in Ireland started about 60,000 years ago and it ended 10,000 years ago.

Glaciers

The climate was very cold during the Ice Age, so cold that the winter snows did not melt from the higher land even in summer. Over a long period, layer upon layer of snow piled up on the land. The weight of snow pressing down compacted the snow underneath into ice. **Glaciers** were formed.

*Sea of ice or Mer de Glace is a **glacier** in the Alps*

Glaciation

Think of glaciers as giant icebergs or frozen rivers sliding over the land. Like a snowplough shifting snow, glaciers moved rocks. As glaciers ploughed their way along, they deepened, widened and straightened the valleys. Glaciers changed the appearance of the landscape completely.

Glaciers shift rock and soil like a giant snowplough moves snow

Erosion by a glacier

Glacial erosion happens in two ways:
- Plucking
- Abrasion.

Plucking happens when the glacier freezes around lumps of cracked and broken rock (see below). Then when the ice moves downhill, rock is plucked away.

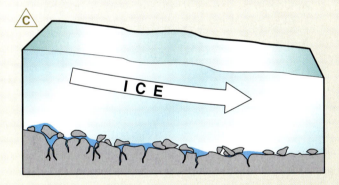

Remember
- Broken and cracked rocks are a result of **freeze-thaw** action (see Chapter 3, Weathering).
- When water freezes in rock it takes up more space. This puts pressure on the rock and it cracks and breaks up.

Abrasion means scraping, using rocks as a tool. The plucked rocks, now stuck into the glacier, dig into and scrape the rock over which the glacier passes. These scrapings are called **striations**.

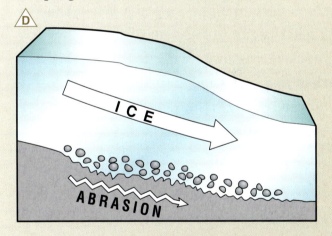

Key points

- **Glaciation** is the effect of ice on the landscape.
- **Glaciers** erode the landscape by freeze-thaw action, **plucking** and **abrasion**.

Landforms of Glacial Erosion

Key words

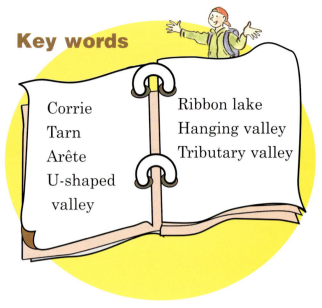

Corrie
Tarn
Arête
U-shaped valley

Ribbon lake
Hanging valley
Tributary valley

Corrie

A **corrie** is formed when ice erodes the land. It is a large hollow in the side of a hill. It has the shape of an armchair, in that it has three steep sides and one open side called a lip. It is the place where glaciers formed.

During the Ice Age, snow and ice piled up in the hollow to make a glacier. When the glacier began to slide out of this hollow it *plucked* rock away with it. It used this rock that was now stuck in it to scrape away (abrade) the ground. The hollow became much deeper.

This deep hollow is called a **corrie**. It is often filled with a lake called a **tarn**. Example: Lake Coomshingaun in the Comeragh Mountains

Sometimes two corries form back to back on a hillside. They are separated by a narrow ridge called an **arête**.

Corrie and lake in the mountains

U-shaped valley

A **U-shaped valley** is a flat-floored, steep sided valley that has been eroded by a glacier.

A valley that has been eroded by a river has a V-shape. When a valley is further eroded by a glacier, the glacier changes this V-shape to a U-shape.

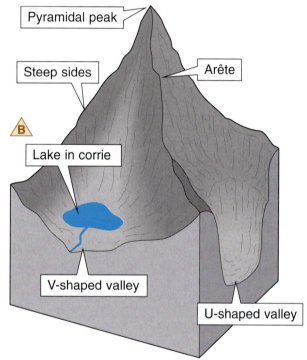

Plucking and abrasion have straightened and deepened the sides and bottom of the valley

During glaciation, a glacier moves down through the valley eroding the rock as it passes along. The glacier moves slowly, which gives it time to cut into the bits of hard rock (spurs) that jut out along the valley. Picture a giant bulldozer shifting all the rock in front of it. Once again, the processes of freeze-thaw action, plucking and abrasion wear down the rocks along the way. In this way, the valley is straightened, widened and deepened by a glacier.

Examples: Glendalough, Co. Wicklow

Ribbon lakes

A glacier can scoop out a hollow in the valley floor. A long deep lake fills the hollow. This is called a **ribbon lake**.

Hanging valley

A **hanging valley** is a valley that hangs over the side of another valley. A waterfall will tumble from this higher valley down into the lower one.

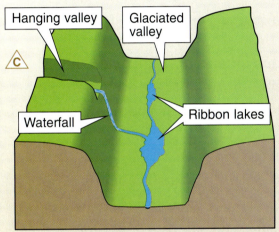

During glaciation, glaciers move down through the valleys. Large glaciers fill the main valleys and smaller ones fill the **tributary valleys**. A tributary valley is a small valley leading into a main valley. The larger glaciers have more power to erode than the smaller ones. This means that the main valley will be deepened more than the smaller tributary valleys. It means that the floor of the main valley will lie far below the floors of the tributary valleys. It also means that the smaller valley will be left **hanging** above the main valley. Then, when the glaciers melt, streams from the tributary valleys will fall down into the main valley. Example: Poll an Eas at Glendalough, Co. Wicklow

Think of two people digging in the ground – one with a big JCB machine and the other with a trowel! They start off at the same level but soon they are at different levels

Pyramidal peaks in the Alps

Key points

- Glaciers erode in highland areas.
- A **corrie** is formed by glacial erosion.
- A **corrie** is the birthplace or starting point of the glacier.
- Glaciated valleys were once river valleys that were deepened, widened and straightened by the ice.
- A **hanging valley** is not deepened as much as the main valley.

Transport of Material by Glaciers

Key words

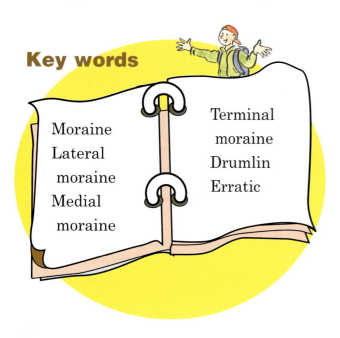

Moraine
Lateral moraine
Medial moraine

Terminal moraine
Drumlin
Erratic

Moraines

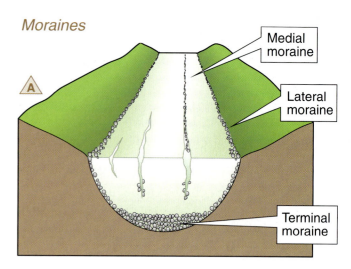

What happens to all the rocks that the glacier has eroded from the sides and the floor of the valley? What happens to all the weathered rocks that fall onto glaciers as they sit in or move down a valley? The glacier carries them away as it moves towards the lowland. This load of material is called **moraine**.

A glacier will deposit the moraine when it reaches lower land. It is warmer there so the ice begins to melt and shrink. When this happens the glacier is unable to shift its load. It drops it on the landscape.

Features of deposition

Moraine is the name given to the material carried along and then dropped by a glacier.

Moraines can look like mountains, hills or large mounds on the landscape. If you cut into them you would see that they are not made up of solid rock. They contain pieces of rocks of all shapes and sizes.

We name different types of moraines according to where the eroded material had rested in the glacier before being deposited.

- There are **lateral moraines**, carried along the sides of a glacier.
- There are **medial moraines** carried along the middle of a glacier.
- There are ground moraines (also called *till*) carried along the bottom of the glacier.
- There are **terminal moraines** pushed in front of a glacier. (See diagram A.)

When the glacier melts, the moraines are dumped and left behind, marking the spot where they were carried in the glacier. Soils develop on these moraines and plants take root.

Example: Upper Lough Bray, Co. Wicklow

Drumlins

A **drumlin** is a small, oval-shaped hill on the landscape. These small hills usually occur in 'swarms' or groups, and from the air, they look like eggs in a basket. They formed towards the end of the Ice Age, when the melting glaciers deposited their load of clay and boulders.

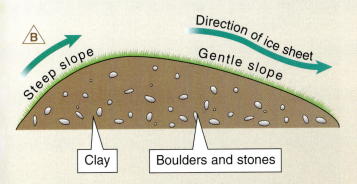

A drumlin is steeper at one end and more gentle at the other. It is usually around 50 metres high and 1km long. It is made of boulder clay, which is a mixture of clay and stones.

The more gentle side of the drumlin points in the direction in which the ice was moving.

Examples: Drumlin belt that stretches from Strangford Lough, Co. Down to Clew Bay, Co. Mayo

Erratics

Erratics are large boulders that were carried by the ice. They were dropped when the glacier lost the power to shift them. These boulders are made of a rock, foreign or different to the rock of the area in which they are found. This tells us they have been moved over a distance by a glacier.

Granite boulders from Connemara are found on top of limestone in the Burren in Co. Clare

Key points

- Glaciers transport large amounts of rock and soil called **moraines**.
- There are different **moraines** depending on where they are carried in a glacier.
- These moraines are deposited on the lowland.
- **Drumlins** are small oval-shaped hills made of boulder clay.
- **Erratics** are large boulders of foreign rock in an area.

Meltwater Features

Key words

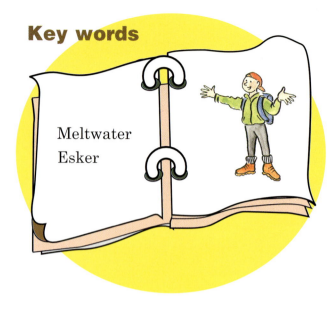

Meltwater
Esker

Esker

An esker is a winding ridge of sand and gravel

When temperatures rose as the Ice Age ended, the glaciers began to melt. Vast amounts of **meltwater** ran out from under the glaciers. These meltwaters carried the eroded material with them. When this material was dropped it formed **eskers**.

Meltwater streams that run out from under a glacier meander just like ordinary streams. They also deposit material on their beds just like rivers do. They continue to flow over these deposits. When the ice has melted you are left with a winding ridge of sand and gravel. This is called an esker. **Eskers** give us sand and gravel for making cement. One esker, the Esker Riada, follows the Dublin to Galway road for some of its journey.

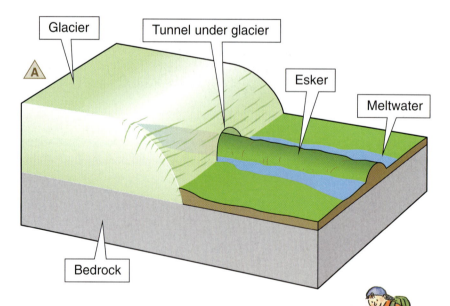

Key points

- As the glaciers melted rivers carrying material flowed out freely.
- These rivers built up ridge-like deposits called **eskers**.
- **Eskers** are valuable sources of building materials.

People and Glaciation

Key words

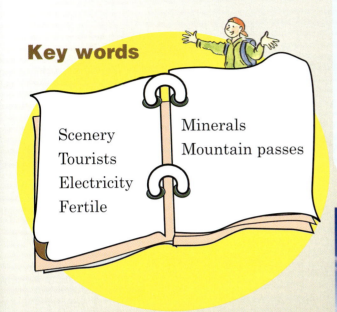

Scenery
Tourists
Electricity
Fertile

Minerals
Mountain passes

Energy

Electricity is produced from the waterfalls and steep valleys. Norway has benefited very much from this cheap electricity.

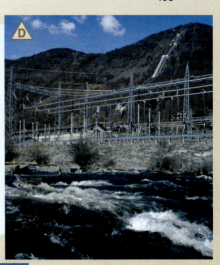

Glaciers shaped landscapes many thousands of years ago. People have made great use of these ice-shaped landscapes.

Farming

Fertile soils often come from the deposits of glaciation. Boulder clays can contain rich **minerals** e.g. The Golden Vale area of Munster.

Tourism

The effect of glaciation produces wonderful **scenery** as in Co. Kerry and Switzerland. This brings in lots of **tourists**, which means more money, jobs and improved services for local people.

Transport

Glaciers cut through mountains as they made their way down slopes to the sea, creating **mountain passes**. Later these passes were used as transport routes and became roads.

Difficulties caused by glaciation

Glaciers stripped soil from some areas leaving a bare landscape. It is difficult to farm in these areas, e.g. the Burren, Co. Clare.

Glacial deposits, like drumlins, often blocked water from draining away. Rainwater then caused waterlogging. This creates problems for farming.

Key points

- Glaciation has brought many benefits for tourism, farming, energy supply and transport.
- Glaciation has also caused problems for farmers.

REVISION EXERCISES

Write the answers in your copybook.

1   In your copybook write out and complete these sentences.
   (a) Glaciers erode by _____.
   (b) The birthplace of a glacier is called a _____.
   (c) Two landforms resulting from glacial erosion are

   _____.

   (d) Two landforms resulting from glacial deposition are

   _____.

2   In your copybook match each letter in column X with the number of its pair in column Y.

X	Y	ANSWER
A Corrie	1 Deposition	A =
B Arête	2 Erosion	B =
C Moraine	3 Erosion	C =
D U-shaped valley	4 Deposition	D =
E Erratic	5 Erosion	E =

3   Write this passage into your copybook and fill in the missing words from the list given.

> **glaciers two decreased ice gravity Kerry
> 10,000 glaciation downslope Wicklow Ice**

From time to time the temperatures of the world _____. This caused the _____Age. There were _____ such times in Ireland. The last one only melted about _____ years ago. It is still not known for definite what caused the temperatures to lower and _____ to form in highland areas.

The effect of ice on the landscape is called _____.
Glaciers are like rivers of _____ . Once they have filled up the hollow in the mountain they move _____. The force of _____ helps them to move.

In Ireland the counties of _____ and _____ have many landforms of glaciation.

4   Name and say where you would find **two** features of glacial erosion in Ireland.

5   Scratches made by the moving glacier on a rock are called s_____.

6   Name **two** features of glacial erosion and **two** features of glacial deposition.

7   Describe **two** benefits of glaciation to people. Write at least three sentences on **each** benefit.

8   Describe **two** bad results of glaciation for humans.

9   In your copybook match each letter in column X with the number of its pair in column Y.

X	Y	ANSWER
A An armchair shaped hollow	1 Drumlin	A =
B A rock that is foreign to the area	2 Corrie	B =
C A ridge of material deposited in front of the glacier	3 Terminal moraine	C =
D An oval shaped hill of boulder clay	4 Erratic	D =
E A ridge separating two corries	5 A hanging valley	E =
F A smaller valley that lies high above the main valley at the end of the Ice Age	6 Arête	F =

10   Draw a clearly labelled diagram showing the difference between a lateral and a terminal moraine. Include a few words to explain both features.

11   The meltwaters from glaciers form features that shape our landscape. One such feature is an esker. Draw and label a diagram showing how eskers are formed.

8 Weather and Climate

Key words

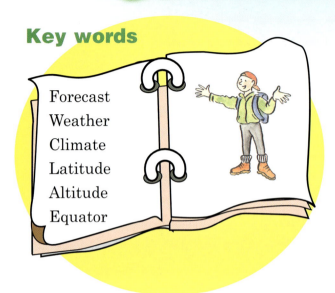

Forecast
Weather
Climate
Latitude
Altitude
Equator

Ⓑ Most areas will be overcast with a risk of showers in the north-west. Cloud will clear and most areas will have sunny spells. Temperatures will reach 18^0 inland but it will be cooler near coasts.

Weather

You have often watched the weather **forecast** on the television. The weather forecast is a prediction of the weather we can expect in the near future. It is based on reports from weather stations and satellite images. You will learn more about this later.

Describe today's **weather**. Is it warm, sunny, windy or wet? Weather is about the air conditions that occur from day to day.

The climate is mild but changeable.

It's hot and dry in summer.

We have a hot wet season and a cooler dry season

Ⓐ It's very cold and dry here.

Equator

It's very hot and very wet here.

My climate has hot days and cold nights.

- ▇ Hot climates
- ▇ Temperate climates
- ▢ Cold climates

Climate

The **climate** of a place is the average weather conditions it gets over a long time (about 30 years).

Different climates

Different parts of the world have different climates. The general picture of climates in the world is of *hot* and *cold* with *temperate* in between.

Why are there different climates?

There are a number of reasons why there are different climates in the world. Let's look at two of them:

- Latitude
- Altitude.

Latitude

Latitude refers to the distance a place is from the **equator**. It is not measured in miles or kilometres. It is measured in degrees. The equator is at 0^0 latitude. The North Pole is at 90^0 latitude. Ireland's latitude is between 51.5^0 and 55^0 north of the equator. Look at the map of the world on page 140 and note the latitude lines.

The hottest places on the earth are near the equator and it gets colder as you move towards the poles. The reason for this is that the sun shines directly on the equator. Further away from the equator, it shines at more of an angle (slant). Slanted rays are cooler.

Think about this. If you sat directly in front of a fire or heater you would be warmer than if you sat at the side. The heat has to travel further to the side and so it will be cooler there.

Altitude – height above sea level

Altitude is the height above sea level. The higher up you go, the cooler it gets. Look at the picture which shows snow at the top of Mt Kilimanjaro in Tanzania, Africa. This is snowcapped though it lies on the **equator**.

Mt Kilimanjaro, Tanzania

The air is thinner higher up and holds less heat. It is like wearing a light or thin jacket. It holds little heat.

Key points

- **Weather** is the daily air conditions.
- **Climate** is the average weather conditions over 30 years.
- **Latitude** influences temperature. Places near the equator are hottest.
- **Altitude**, the height of a place above sea level, also affects temperatures.

The Role of the Sea

Key words

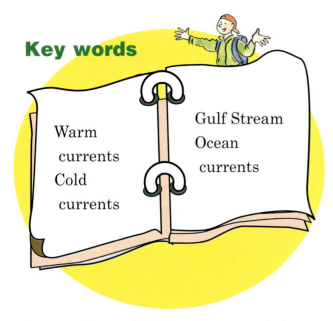

Warm currents

Cold currents

Gulf Stream

Ocean currents

The sea affects the temperatures of places near the coasts. In summer, places near the coast are cooler than places that are inland.

In winter it is the opposite. Places near the coast are warmer in winter than places that are inland. Look at the map.

Temperature °C	Valentia	Berlin	Warsaw	Moscow
January	7°C	–1°C	–3°C	–15°C
July	15°C	18°C	19°C	20°C

A cool breeze will blow in from the sea in summer keeping places near the coast cool.

A warm breeze will blow in from the sea in winter keeping coasts warm. Away from the sea it can get very cold in winter.

In Ireland, winter frosts are much more common inland than on the coasts.

Ocean currents

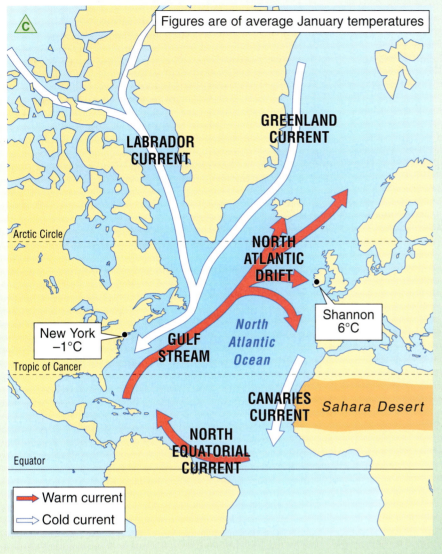

Figures are of average January temperatures

C

GREENLAND
CURRENT

LABRADOR
CURRENT

Arctic Circle

NORTH
ATLANTIC
DRIFT

Shannon
6°C

New York
−1°C

GULF
STREAM

North
Atlantic
Ocean

Tropic of Cancer

CANARIES
CURRENT   Sahara Desert

NORTH
EQUATORIAL
CURRENT

Equator

→ Warm current
⇨ Cold current

There are **warm currents** and **cold currents** in the sea.

A warm current will heat the air above it. This warmed air will blow in over the land raising the temperatures there. There is a warm current called the **Gulf Stream** (or North Atlantic Drift) flowing through the Atlantic Ocean towards Ireland. It heats the south-west wind, which then blows over Ireland as a mild wind. Without the influence of the Gulf Stream, Ireland's temperatures would be lower.

In eastern Canada, the cold Labrador current cools the air above it. The cooled wind blows in over the land bringing temperatures down very low in winter.

D

Snow bound St Lawrence Seaway in winter

Key points

- The sea has an influence on temperatures.
- Places close to the sea are warmer in winter than places further from the sea.
- Places close to the sea are cooler in summer than places further from the sea.
- **Ocean currents** affect the temperature of the air above them.

The Influence of Wind

Key words

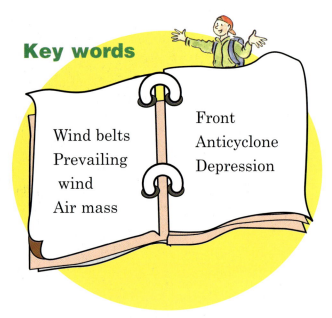

Wind belts
Prevailing
wind
Air mass

Front
Anticyclone
Depression

Wind is nature's way of trying to even out temperatures on the earth. If all places got the same amount of heat from the sun, we would not have a wind. Winds can be warm or cool. A warm wind will raise the temperature of a place. A cool wind will lower temperature.

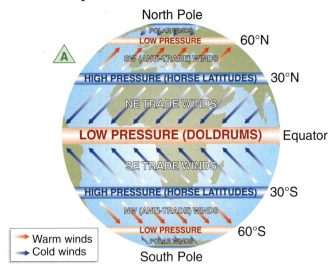

Diagram **A** shows the **wind belts** of the world. You will notice the following:

- Air blows from the areas marked high pressure to the areas marked low pressure.
- Air is deflected (moved off a straight line) to the right of its path in the northern hemisphere and to the left in the southern hemisphere.

- Winds are named according to the direction *from* which they blow. In the Northern Hemisphere:
- Winds from the north lower temperatures.
- Winds from the south raise temperatures.

Winds over Ireland

The **prevailing wind** over Ireland is the south-west wind. Prevailing wind means it is the usual wind. It is called the south-west wind because it blows from the south west. The south-west wind is mild because it is blowing from the direction of the equator where temperatures are higher.

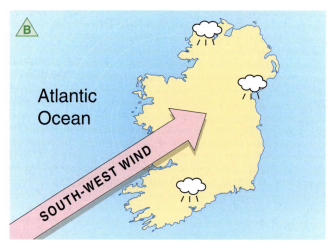

The south-west wind carries a lot of moisture with it because it has been blowing over the sea picking up moisture as it goes along. This wind brings a lot of cloud and rain to Ireland. The west of Ireland gets more cloud and rain than the east because the south-west winds drop much of their rain when they first move in over the land.

Sometimes it happens in Ireland that the wind blows from the north. The north wind is a cold wind and in winter it could bring snow.

Air masses

It can happen that warm air from the south (a tropical **air mass**) can meet cold air from the north (a polar air mass). At the meeting place of the two air masses, you get a **front**. There are warm fronts and cold fronts. Both bring rain and changeable air conditions. You will have seen this on a weather forecast map.

Ireland is at a place on the earth, the mid-latitudes, where you have the meeting of cold and warm air. This is what gives us the changeable conditions we know so well. You will learn more about this when we look at rainfall.

NORTH WIND

Air pressure

Air has a weight. It can be heavy or light.

When air has a high pressure (H) there will be few or no clouds. In summer this means warm temperatures, slack winds and descending air.

When air has a low pressure (L) there will be clouds. This can mean lower temperatures, wet changeable weather, strong winds and ascending (rising up into the sky) air.

In Ireland high pressure means a dry settled spell of weather. A high-pressure system is called an **anticyclone**. A low-pressure area is called a **depression** or cyclone.

Key points

- Wind can raise or lower temperatures.
- Winds are named after the direction from which they blow.
- A **front** is the boundary between two **air masses**.
- **Depressions** mean wet changeable weather conditions.
- An **anticyclone** in a weather system means dry settled weather.

The Water Cycle

Key words

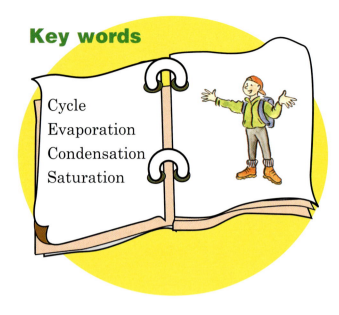

Cycle
Evaporation
Condensation
Saturation

Evaporation: This is the means by which water turns into a gas called water vapour.
Condensation: This is the means by which water vapour turns back into liquid drops.
Saturation: This is when something, for example clouds, is full of water.

Water is passed around between the oceans, the clouds and the land. It goes around and around in a **cycle**. Look at diagram A of the water cycle.

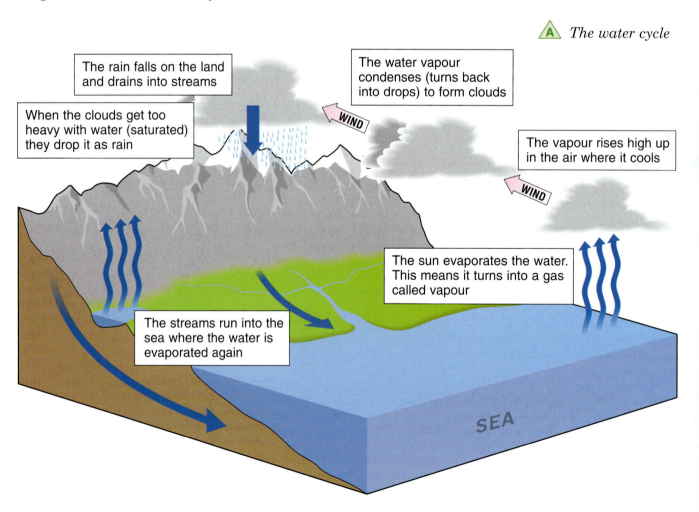

A   *The water cycle*

The rain falls on the land and drains into streams

The water vapour condenses (turns back into drops) to form clouds

When the clouds get too heavy with water (saturated) they drop it as rain

WIND

The vapour rises high up in the air where it cools

WIND

The sun evaporates the water. This means it turns into a gas called vapour

The streams run into the sea where the water is evaporated again

SEA

Clouds

Not all clouds are the same. Look at the pictures showing the different clouds. The type of clouds in the sky can give us an idea of the type of weather to expect.

Cumulus-nimbus clouds bring rain, lightening and thunderstorms

Stratus clouds are like a blanket blocking out any trace of blue in the sky. They bring continuous spells of rain

Cumulus clouds bring heavy bursts of rain

Cirrus clouds are very high in the air. No rain is expected

Key points

- The water cycle involves **evaporation**, **condensation**, rainfall and run-off.
- Different clouds are linked to different weather conditions.

Rainfall

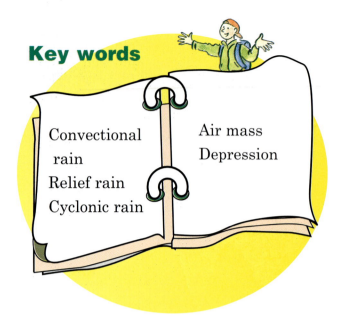

Key words

Convectional rain
Relief rain
Cyclonic rain

Air mass
Depression

To understand how convectional rain is formed, think of what happens when you boil a kettle of water. The water is heated and turns into steam or water vapour. When the steam hits the cold ceiling or windows, it is cooled. It turns back into water drops that fall back down to the ground.

Now think of conditions on a very hot day. The air near the ground is heated. Water turns into vapour. It rises up and cools in the cold air higher up. It turns into water drops and falls as rain. This is called convectional rain.

The word convection has to do with currents of warmed air rising up.

Rain is formed when air is cooled. Because air can be cooled for different reasons, there are different types of rainfall. These are:

- **Convectional rain**
- **Relief rain**
- **Cyclonic rain**.

Convectional rain

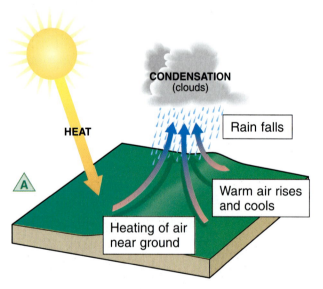

CONDENSATION (clouds)

HEAT

Rain falls

Warm air rises and cools

Heating of air near ground

A

Convectional rain comes in heavy downpours. It is common in very hot countries like Nigeria and Brazil. Sometimes in Ireland we have this type of rainfall. This will happen in summer when temperatures become really warm and there are thunderstorms and heavy downpours.

Relief rain

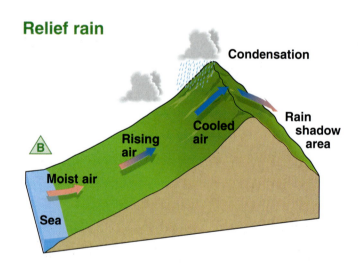

Condensation

Rain shadow area

Cooled air

Rising air

Moist air

Sea

B

Relief rain is the kind of rain that falls over high ground. It gets its name from the word **relief** which has to do with the height of the land. This type of rain is common in Ireland especially in the mountainous areas of the west.

Rainfall map of Ireland

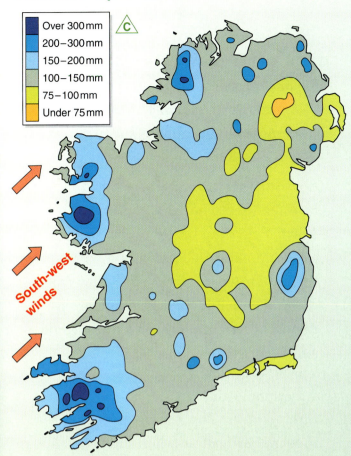

The rainfall map of Ireland shows that most rain falls on mountains. The mountains along the west coast get the highest amounts of rain because the prevailing winds blow in from the south west.

Picture a wind carrying a lot of moisture. It blows in from the sea and it meets the mountains. It cannot pass through the mountain so it is forced to rise up over the top. As it rises up towards the higher ground it is cooled down. The water condenses and falls as rain on the mountain. As the wind blows down over the other side it is warmed again. This side of the mountain will be drier. This is called the *rain shadow* side.

The third type of rainfall is called **cyclonic** or depression rain. This type of rain is very common in Ireland.

Remember the key point that for rain to fall, air must be cooled. Picture two air masses meeting. One **air mass** is warm; the other air mass is cold. The warm air mass is cooled when it hits the cold air mass. The moisture will condense, form clouds and fall as rain.

Cyclonic rain

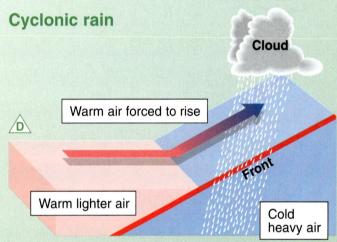

A weather system, called a **depression,** forms at fronts. A front is the meeting place of two air masses. Depressions are low-pressure weather systems that bring changeable and damp conditions to Ireland.

Key points

- There are three types of rain: **convectional**, **relief** and **cyclonic**.
- **Relief** rain brings higher rainfall to the mountains along the west coast.
- **Convectional rain** happens when warm moist air rises and cools.
- **Cyclonic rain** happens when two air masses meet at a front.

Ireland: Weather and Climate

Key words

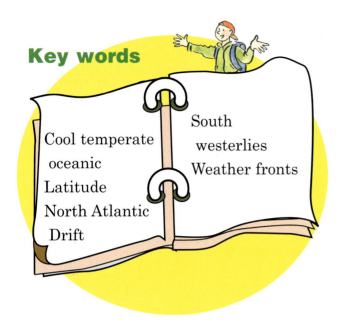

Cool temperate oceanic

Latitude

North Atlantic Drift

South westerlies

Weather fronts

Ireland's climate is called **cool temperate oceanic**. Temperate means that temperatures are neither very hot in summer nor very cold in winter. Oceanic suggests that the sea has an influence on our weather conditions. Ireland's weather is *changeable*.

The main factors influencing Ireland's climate are:

- **Latitude** – Ireland is almost mid-way between the equator and the poles. This means temperatures are moderate, i.e. neither very hot nor very cold.

- The sea greatly influences our climate. As we are an island we are surrounded by water. This brings cooler air in summer and warmer air in winter.

- A warm current, the **North Atlantic Drift**, increases the temperature of the sea off our west coast.

- Our prevailing winds, the **south westerlies**, which come from warmer latitudes and pass over this warm current, bring mild and wet conditions.

- **Weather fronts**, the place where cold air and warm air masses meet, are found over Ireland. This makes the climate changeable and wet.

Cool summers: Because Ireland's latitude (51° to 55°N) is closer to the North Pole than to the equator it has cool summers. Our prevailing winds are the **south westerlies**. They blow from the cooler sea during the summer so temperatures are lower than you could expect at this latitude.

Mild winters: The sea has kept some of its summer heat so when the prevailing winds, the south westerlies, blow in from the Atlantic they carry this warmer air. This gives Ireland mild winters.

Rainfall: Rain falls throughout the year. The south-west winds pick up moisture as they cross the Atlantic, bringing plenty of rain to Ireland.

A What will I wear? The weather is so changeable! That's what keeps us on our toes.

Graph 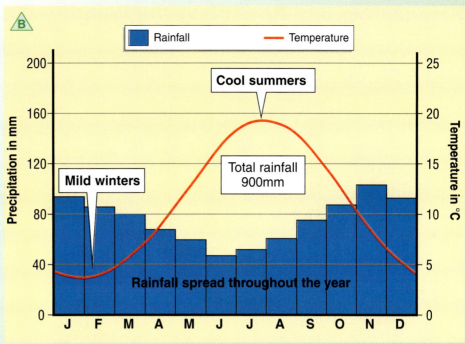 shows the weather statistics (figures) for Ireland. The bars show the rainfall amounts for each month. You can read the amount by following the numbers on the left side of the graph. The red line shows the temperatures for each month. You can read the numbers by following the figures on the right side of the graph.

Ireland's prevailing wind

SW ⇒ South-west winds

Monday   Tuesday

Key points

● Ireland's climate is described as a **cool temperate oceanic** type.

Forecasting and Predicting the Weather

Key words

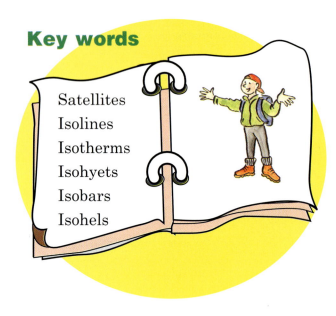

Satellites
Isolines
Isotherms
Isohyets
Isobars
Isohels

Weather stations in Ireland

Readings of weather conditions are taken at weather stations all over the country. Some information is beamed from **satellites**. The gathering of this information makes it possible to predict weather conditions for the coming days.

Once information is gathered at the Central Forecasting Office in Glasnevin, it is then plotted on a map. The map uses lines called **isolines** to show the information. It allows us to see similarities and differences in weather conditions between areas and from day to day.

Wind direction Depression

Wind direction Anticyclone

Different weather charts

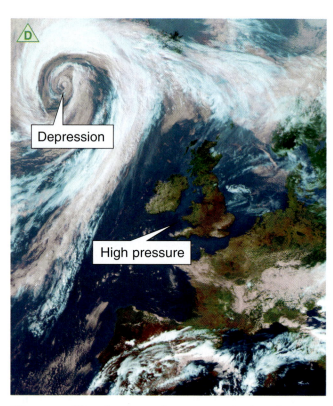

The satellite image shows an anticyclone or high pressure over Ireland. There is a depression to the north west of the country. Knowing this allows the forecasters to predict likely weather conditions.

Weather instruments

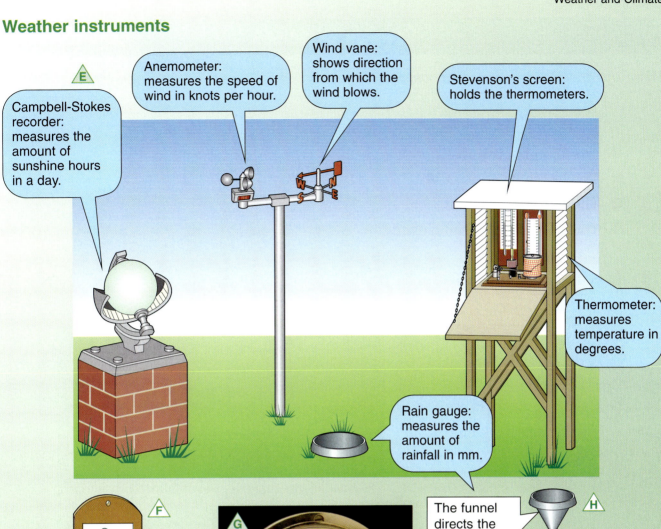

Campbell-Stokes recorder: measures the amount of sunshine hours in a day.

Anemometer: measures the speed of wind in knots per hour.

Wind vane: shows direction from which the wind blows.

Stevenson's screen: holds the thermometers.

Thermometer: measures temperature in degrees.

Rain gauge: measures the amount of rainfall in mm.

The funnel directs the water into the measuring cylinder

The measuring cylinder is marked in millimetres

The outer cylinder

Barometer: measures the pressure of the air (hPa).

Thermometer: as temperature rises, the mercury in the tube rises.

Key points

- The weather forecast tells what the weather will be like over the coming few days.
- Information is collected at weather stations.
- All the information is put together at the Central Forecasting Office and this information is used to make a weather forecast.

Summary

Element	Instrument	Unit of Measurement	Maps
Temperature	Thermometer	Degrees Celcius or Centigrade	Isotherm
Rainfall	Rain gauge	Millimetres	Isohyets
Air pressure	Barometer	Millibar	Isobars
Sunshine	Campbell-Stokes recorder	Hels or hours	Isohels
Wind speed	Anemometer	Knots	Number
Wind direction	Wind vane	Directions	Arrow

REVISION EXERCISES

Write the answers in your copybook.

1 Which ocean current brings warm water past the coast of Ireland?
- The Canary current
- The North Atlantic Drift
- The Labrador current
- The North Equatorial current

2 In your copybook match each letter in column X with the number of its pair in column Y.

X	Y	ANSWER
A Thermometer	1 Rainfall	A =
B Wind vane	2 Wind speed	B =
C Anemometer	3 Sunshine hours	C =
D Campbell Stokes recorder	4 Temperature	D =
E Rain gauge	5 Wind direction	E =

3 Draw and carefully label the following instruments in your copybook:
- A rain gauge
- An anemometer
- A thermometer

4 Draw three diagrams to show the different types of rain. For each diagram name one area where that type of rain falls.

5 Write out a weather forecast for one day of this week. Use the TV weather forecast to help you. Include a diagram showing the isobars.

6 Copy the text below into your copybook and fill in the missing words.
A_____ measures the amount of pressure in the atmosphere.
A depression is an area of low pressure that brings _____ weather. An anticyclone brings _____ weather.

7 What is an ocean current? Write two sentences describing how a warm ocean current affects the climate of Ireland.

8 Four different sources of our weather information are:

T_____

R_____

N_____

Central F_____ Office, Glasnevin

9 Write out the temperature for each day of the week. You may have to use the daily newspaper to do this. Calculate the mean/average weekly temperature.

10 Draw a Stevenson screen in your copybook. Clearly show the main characteristics (hint: colour, design of sides). Which instruments lie in this box in a weather station?

11 The following are the temperatures for an eight-hour period. What is the range of temperatures? (This means the difference between the highest and the lowest figure.)

9 am − 16°
10 am − 16°
11 am − 17°
12 am − 19°
1 pm − 19°
2 pm − 21°
3 pm − 24°
4 pm − 26°

12 Look at the images. Many people depend on the weather for their livelihoods. Name two of these and explain why the weather information is so important to them.

13 Look at the map that shows the weather stations in Ireland on page 94. Name four weather stations and state what county they are in.

9 Climates of the World

Key words

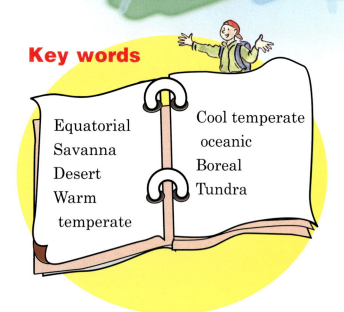

Equatorial
Savanna
Desert
Warm temperate

Cool temperate
oceanic
Boreal
Tundra

World Climatic Divisions

Climate refers to the average weather conditions that a place has over about 30 years. Climatic conditions influence the plant life, animal life and human activities in different parts of the world.

When studying the climate of a place you will need to know about:

- The temperatures, air pressure, rainfall, wind, sunshine conditions and what causes these to occur.
- The natural vegetation, because climate has a huge influence on what grows in a region.
- Human activities in that region, because climate influences the way people live, the clothes they wear and the food they eat.

Areas that have similarities in climatic conditions, animal and plant life, along with common human activities are called **natural regions**.

Types of climates

There are a number of climate types in the world. They broadly fall into three groups:

- Hot climates
- Temperate climates
- Cold climates.

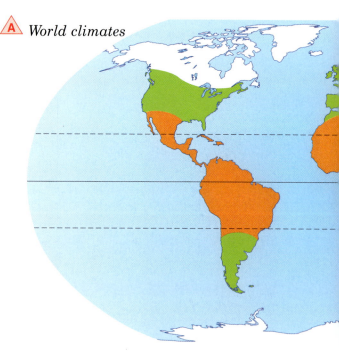

A *World climates*

B *Hot climates – subdivisions*

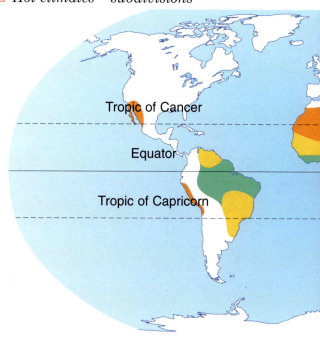

Tropic of Cancer

Equator

Tropic of Capricorn

As you move further away from the equator the climate changes from hot to temperate to cold. Look at map showing these three groups.

The hot, temperate and cold climates are further divided into different types of climate. Look at the chart.

Climate	Subdivisions	Latitude
Hot climates	**Equatorial** **Savanna** (tropical grasslands) **Desert**	Between 30° north and south of the equator
Temperate climates	**Warm temperate** (Mediterranean) **Cool temperate oceanic**	Between 30° and 60° north and south of the equator
Cold climates	**Boreal** **Tundra**	North of 60°

You will be looking in more detail at these climates: **Desert**, Mediterranean and **Tundra**.

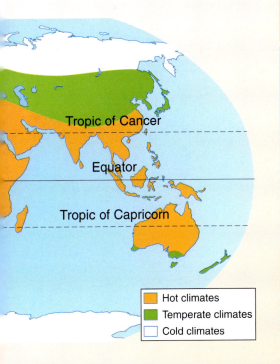

Tropic of Cancer

Equator

Tropic of Capricorn

- Hot climates
- Temperate climates
- Cold climates

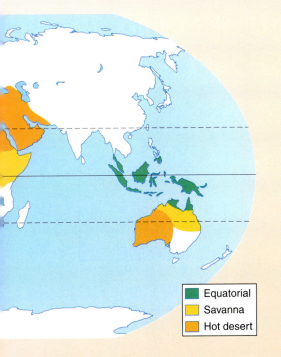

- Equatorial
- Savanna
- Hot desert

Key points

- Climate is the average weather conditions over about 30 years.
- The three broad divisions of climate are **hot**, **temperate** and **cold**.
- **Equatorial**, **savanna** and hot deserts are examples of climates in the hot climate group.

10 Hot Desert Climate

Key words

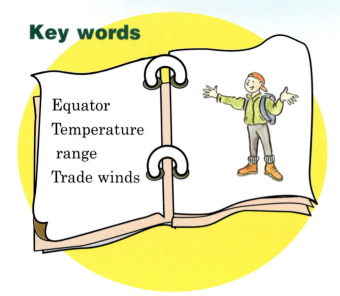

Equator
Temperature
  range
Trade winds

Which places on the earth have a desert climate?

Look at the map showing the hot deserts of the world. They lie between 15–30° north and south of the **equator**.

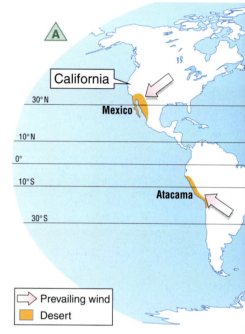

California
30°N
Mexico
10°N
0°
10°S
Atacama
30°S

⇨ Prevailing wind
▦ Desert

Climate of the deserts

The hot deserts of the world are very hot during the summer months. They are cooler in the 'winter' months. (Winter in the deserts means the cooler season. Deserts do not have cold winters.)

The deserts are very dry but they do get some rain. Rainfall amounts are very small – less than 100 mm a year. (Compare this with Dublin, it gets about 800 mm of rainfall a year.)

Look at graph **B** that shows the climate figures for a desert.

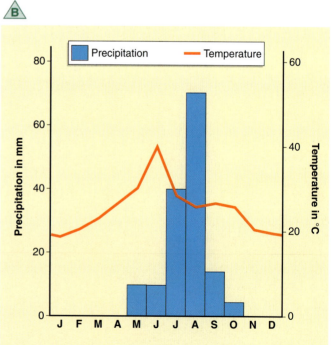

A weather graph showing temperature and precipitation in Khartoum, Sudan

Hot summers in deserts

Deserts are hot because:

- They are near the **equator**. They get direct or nearly direct rays of the sun during the summer months. Direct rays are very hot.
- Winds blowing in on the deserts blow over warm land bringing heat to the deserts.
- Temperatures become very warm in the daytime because there are no clouds to block out the heat.

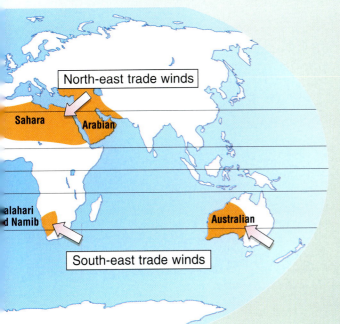

Cooler temperatures in winter months

Deserts are cooler in the winter months. The rays of the sun strike the earth at a lower angle and are weaker at this time of the year.

Low rainfall in the deserts

There is little or no rainfall in deserts because they get very dry winds, called **trade winds**.

The winds are travelling over land towards the equator. They do not pick up moisture. No clouds can form. Because clouds are not formed, there will be no rain.

Day and night temperatures in the desert

There is a huge difference between the temperature of deserts during the day and the temperatures at night. Days are very hot. Nights are very cold. The difference can be as much as 40°C. We call this a **temperature range** of 40°C.

The reason for this difference is that there is no cloud cover over the deserts.

During the day the air in the desert becomes very hot. There are no clouds to block the heat of the sun. Temperatures can rise to more than 40°C. You would find this unbearably hot.

At night the temperatures can fall to freezing point or below. This is because there are no clouds. The heat escapes back to space because there are no clouds to keep the heat in.

Daytime   *Night-time*

Key points

- Deserts lie between 15 and 30° north and south of the **equator**.
- Deserts are very hot and dry.
- Daytime in the deserts is very hot but night-time is much colder.
- Dry winds called the **trade winds** blow over the deserts.

101

Life in the Desert

Key words

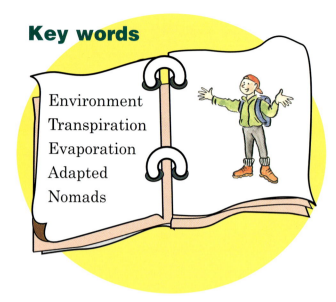

Environment
Transpiration
Evaporation
Adapted
Nomads

To survive in the hot, dry conditions of the deserts, plants must be able to find water and store it. They must also be able to survive the very cold nights.

Look at the picture to learn about the various ways plants adapt to (fit in with) the **environment** of the desert.

A desert landscape

Some plants have spiky leaves – they do not lose water through **transpiration.**

Long tap roots to tap into water deep in the ground.

Thick, waxy skin to lessen the amount of water lost through **evaporation.**

Plants are well spread out so there are not too many plants trying to draw on the water in a small area.

Spongy texture and fleshy stems can store water.

Animals in the deserts

Animals in the deserts have to be able to survive the hot, dry conditions. Animals, like plants, have **adapted** to these conditions. Take a look at how a camel is built.

People in the deserts

Some people have managed to make a living in the desert in spite of the harsh climate. Like animals and plants, they have adapted to the conditions.

Thick lips to deal with the prickly surface of many desert plants.

Long eyelashes and a second eyelid to deal with the sand blowing around during sand storms in the deserts.

A fatty hump from which to draw moisture during dry conditions.

Padded feet to ease walking over a sandy surface.

Nomads

People of the desert are **nomads**. They travel in search of food and water for their herds of camels, goats or cattle. One tribe that does this is the Tuareg tribe of North Africa. They do not stay too long in any one place. They are careful not to overgraze the land. They must leave some covering of vegetation for seeds to take root again. They move on and leave the land to rest before they return.

Their clothes show how they have adapted to the fact that days are very hot but nights can be freezing.

White garments keep desert dwellers cooler. Light colours reflect the rays of the sun. Being well wrapped up means no danger of sunburn

Key points

- Plants, animals and humans have **adapted** to the hot dry conditions of the deserts.

Humans and Deserts

Key words

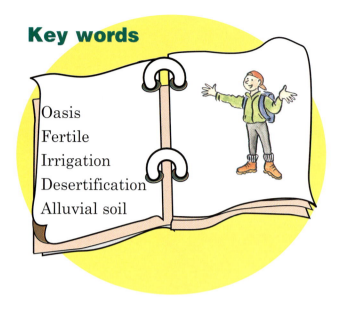

Oasis
Fertile
Irrigation
Desertification
Alluvial soil

Irrigation

On a small scale when you water your plants in your garden or flowerpots, you are irrigating them!

Many parts of the deserts have been made fertile by bringing water to them by pipeline. An example of this is in Egypt in Africa. Water from the River Nile is carried over long distances and used to irrigate the land. (Look at the case study opposite.)

Oasis

Even in the desert there is water. It can appear at or near the surface. This makes an **oasis** or **fertile** spot. An oasis is fertile because plants can reach water. Sometimes wells are sunk and the water is drawn up and spread on the land. With the high amounts of sunshine, warm temperatures and a water supply, crops such as oranges, dates, figs, vines and tobacco can grow very well. The artificial watering of the land is called **irrigation**.

An oasis in the desert

Desertification

Irrigating the deserts shows how people can change deserts to fertile places. However, it has also happened that human activities have had the opposite effect. People have made deserts. This is called **desertification**. This is happening in the Sahel region of North Africa.

You will learn about this in Chapter 13, Soils.

Case Study: The Aswan Dam Project on the River Nile

The Aswan Dam is built on the River Nile in Egypt. It is the world's highest dam and it holds back the water of the River Nile to make the world's largest man-made lake called Lake Nasser. The lake is 500km long, 22km wide and 90 metres deep.

Building the dam started in 1960. It took ten years to complete. The river was held back behind the wall of the dam and a huge area was then flooded to make the large lake.

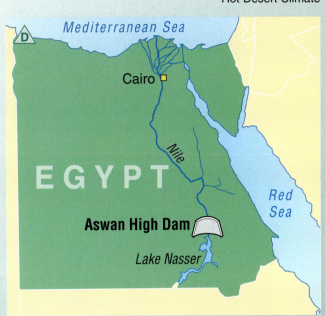

The Aswan Dam is built on the River Nile

Good effects

The benefits (good effects) of damming back the water are:

- Water can be used to irrigate the deserts of Egypt and the Sudan.
- Crops can be grown to feed the people.
- Trapping water behind the dam also means the force of the water can be used to make electricity for industry and for homes. Every city, town and village in Egypt now has electrical power.

Bad effects

There were some disadvantages (bad effects) of the project:

- People had to move from the area where the dam and lake were created.
- The Nubian people of Egypt had to leave their homeland and settle on the newly irrigated land.
- Ancient tombs and temples had to be moved before the area was drowned under the man-made lake, Lake Nasser.
- The Nile used to flood every year. The floodwaters carried **alluvial soil** to the land making it fertile (rich). When the lake was made and the waters held back, the rich soils were no longer spread across the land.

The Aswan Dam in Egypt

Key points

- People have **irrigated** the deserts and made them **fertile** enough to grow crops.
- Because of human activities, the deserts are spreading. This is called **desertification**.

Revision Exercises

Write the answers in your copybook.

1 Which of the following statements best describes the hot desert type climate?
- Short summers and long winters
- Hot dry summers, warm moist winters
- Heavy rainfall throughout the year
- Hot dry summers, cold winters

2 The hot deserts of the world are to be found:
- Only between 15–30° north of the equator
- 30–40° north and south of the equator
- Between 0–8° north and south of the equator
- Between 15–30° north and south of the equator

3 Where is the region known as the Sahel found?
- Southern Europe
- North Africa
- California
- India

4 An example of a country which has a hot desert type climate is:
- Greece
- Netherlands
- Mali (North Africa)
- Spain

5 Name three hot deserts of the world.

6 In your copybook match each word in column X with the number of its pair in column Y.

X	Y	ANSWER
A Desertification	1 Animal of the desert	A =
B Camel	2 A place where there is water	B =
C Tuareg	3 The spread of the desert	C =
D Oasis	4 Artificial watering	D =
E Irrigation	5 A tribe living in the desert	E =

106

7   Explain why during the day it is very hot in the desert, but during the night it is very cold.

8   Show, using carefully labelled diagrams, how one plant and one animal have adapted to the hot conditions found in the desert.

9   Write this passage into your copybook and fill in the missing words from the list given.

> **clouds   nomads   Nile   15°   Aswan Dam**
> **cold   camel   irrigation   south   oasis**

The hot deserts of the world lie between _____ and 30° north and _____ of the equator. During the day it is very hot but during the night it is very _____. This is because there are no _____. Throughout time the people who live here moved from place to place and were called _____. Their main means of transport was the _____. Whenever they came to a watering hole called an _____ they stayed there for a while. Water is scarce in the deserts but whenever possible _____ schemes are built to supply water. One such scheme is called the _____ _____. It was built on the River _____ in Egypt.

10   Write in your copybook the correct answer for each of the statements below.

     (a) The main winds blowing over the hot deserts are the **trade/south westerlies**.

     (b) The main dam on the River Nile is called the **Hoover/Aswan Dam**.

     (c) A plant that survives in the desert is a **bamboo tree/cactus**.

11   Explain, using **one** example, what is meant by the word 'desertification'.

12   The Aswan Dam has changed the life of many people along the banks of the River Nile. Describe two benefits and two bad effects of the project for local people.

Key words

Warm temperate
Mediterranean
Drought

Prevailing winds
Moist

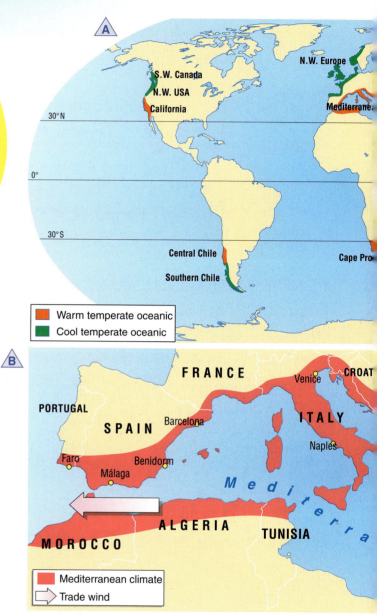

There are three broad climate groups – hot, temperate and cold. Look back at map **A** on pages 98 and 99. Now look at map **A** opposite showing some temperate climates.

Let's look at **warm temperate** or **Mediterranean** type climates. The word temperate means there are no extremes of temperature. These areas have a warm climate.

Which places have a Mediterranean type climate?

Look at the map showing the areas of the world that have a **warm temperate** climate. They lie between 30° and 40° north and south of the equator. It is the climate you find in the south of France, Spain, Italy and Greece. These countries all lie around the Mediterranean Sea. That is why this climate is also called **Mediterranean** type.

Mediterranean type climate – temperatures and rainfall

Summers are very warm and dry. Temperatures average 23°C and there is a summer drought. A **drought** means a long spell without rain.

Winters are mild and **moist**. Temperatures are about 10°C and rainfall averages 800 mm per year. Look at the graph in diagram **C**.

Warm dry summers

The climate conditions in these areas in summer are similar to conditions in the deserts. Though not as hot as deserts, they are still very warm and very dry in summer.

Summers are very warm and dry because:

● These lands are not too far from the equator. In the summer the sun is high in the sky so the rays are very warm. If you have ever visited Spain or the South of France, you will have noticed this.

● The **prevailing** (usual) **winds** are the trade winds. These are warm winds. These warm winds keep temperatures high during the summer.

● The winds in summer are dry winds. This is because they do not come from the sea, but blow from the land. They do not carry cloud and rain to these areas.

Mild wet winters

Winters are cooler in the Mediterranean regions because the sun is lower in the sky. You will remember that more slanted rays are cooler. But because these areas are not too far from the equator it is rarely cold, even in winter.

These regions get rain in the winter because the **prevailing winds** now blow from the sea. As in the case of Ireland, winds from the sea bring cloud and rain.

If you look at map **B** closely, you will notice that the areas that have a Mediterranean type climate are on the west coast of the continents. In the winter they are in the path of the winds blowing from the sea.

Key points

● Places with a **Mediterranean** type climate have very warm, dry summers and mild **moist** winters.
● Winters are moist because the winds blow from the sea.
● Summers are dry as **prevailing winds** blow across the land.

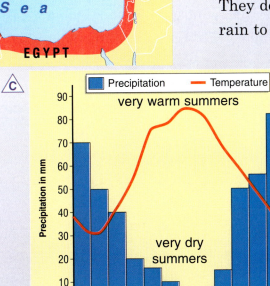

Mediterranean climate

Living in Mediterranean Lands

Key words

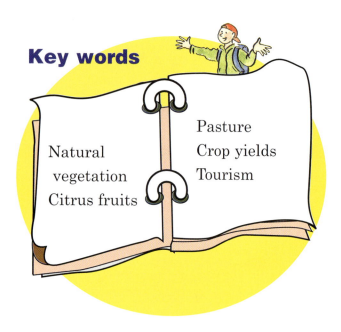

Natural vegetation
Citrus fruits

Pasture
Crop yields
Tourism

The most common tree in the Mediterranean regions is the olive tree. With its waxy leaves and long roots it survives well here. If you have ever been to Spain or Italy on holiday you will know that olive oil is used widely in cooking. Other fruit trees include lemons and oranges. These are known as **citrus fruits**. Vines (grapes) are grown in vineyards all over these lands. These crops grow well in the warm, dry sunny climate.

Natural vegetation means plants that grow naturally in the wild. They do not need to be planted by humans. The natural vegetation in any region has adapted to the climatic conditions in that region.

Vineyards in Mediterranean lands

The natural vegetation of Mediterranean lands can survive the long, summer drought. Typical plants found here include cactus, cork oak, cedar and cypress trees. Many have waxy leaves so the plants do not lose their moisture in the long hot summers. They have long roots to reach down to the water stored in the rocks underground. (This is called the groundwater.) Some have a spongy texture so they can absorb water and store it.

Animal life

The poor **pasture** (grass) of the Mediterranean regions makes it difficult for animals such as cows to survive well. Sheep, goats and pigs are common here. They have adapted better to the vegetation and climate of these areas. Sheep have steady feet. This allows them to move easily in mountainous areas. They can survive on the poor pasture that grows where little rain falls.

In this way, plants have adapted to the very warm, dry conditions in these Mediterranean lands.

Sheep and goats can survive on poor pasture

Human activities in Mediterranean regions

Farming

Farming has always been one of the main activities in these regions. However, the shortage of water in summer and the lack of money available for improvements made farming difficult in the past. **Crop yields** (amount that grew) were small. Farmers made little money.

Nowadays, with the help of money from sources like the EU, irrigation schemes have been set up. Bringing water by pipeline to the land has made it possible to grow other crops apart from the usual vines and olive trees. Vegetables and flowers are grown for sale in the local tourist resorts. They are also sent to markets in the cities of Europe further north. The farmers are better off now than in the past.

Tourism in Mediterranean lands

From the 1950s, the Mediterranean areas became important for **tourism**. People came from the colder, industrial cities further north. The guaranteed sunshine and warm temperatures attracted tourists. Leisure activities such as swimming, sunbathing, windsurfing and jet-skiing were developed. Apartments and holiday villas were built and road, rail and air connections were improved.

Crops growing on irrigated land

Today over 100 million tourists visit these areas each year. Places such as Spain, the South of France, Italy and Greece are the most popular areas in Europe for holidays. **Tourism** is hugely important in these areas because:

- Jobs have been created in the building of hotels and roads.
- Jobs have been created in hotels and restaurants.
- Services, such as roads, have been improved in the region.
- Tourism has created a market for crops grown in the area.

Key points

- Olives and vines have adapted to the hot, dry conditions.
- Irrigating the land means that more vegetables and fruits can be grown.
- The warm, sunny climate attracts millions of tourists each year.
- **Tourism** brings jobs and money to these areas.

REVISION EXERCISES

Write the answers in your copybook.

1   Which of the following best describes the Mediterranean climate?
   ● Very hot summers, cool winters, rainfall in summer
   ● Very hot, very wet, no difference between seasons
   ● Hot summers, warm winter, rainfall in winter
   ● Warm summers, mild winters, rainfall throughout the year

2   The word drought means:
   ● Spiky   ● Lack of rain
   ● Sunny   ● Artificial watering

3   Write out the correct answer for each of the statements below:
   (a) The Mediterranean type climate is found in **Greece/Norway**.
   (b) The typical tree crop found here is **pine/olive**.
   (c) The main animals farmed are **sheep/cows**.

4   Name two areas that have a Mediterranean type climate.

5   Give three reasons why over 100 million people visit Mediterranean regions every year.

6   Show, using a diagram, how one plant found in Mediterranean regions has adapted to the climate.

7   The information in the table and pie chart below shows the percentage (%) areas of different fruits on a Mediterranean farm. Copy the pie chart into your copybook and write the names of the crops in the correct spaces, using the information in the table.

Crop	% Area
Grapes	50
Peaches	25
Plums	10
Apples	10
Tomatoes	5

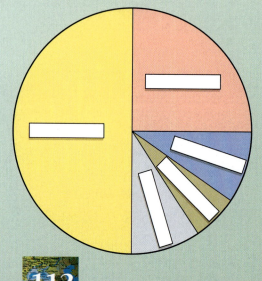

8   Farmers now use irrigation to increase the amount of crops that
    grow (the yield).
    (a) Explain how irrigation makes this possible.
    (b) Name three food crops grown in these areas.

9   Look at the map of the Mediterranean areas and name:
    ● Two areas of Europe
    ● Two areas of North Africa

10 Explain the following terms:
    (a) Drought
    (b) Natural vegetation
    (c) Irrigation schemes

11 Copy this passage into your copybook. Then using the words from
    the box below fill in the blanks.

> **lemons   40°   South of Spain   olive   dry**
> **oranges   drought   Mediterranean**

The Mediterranean climate is found between 30° and _____
north and south of the equator. It is usually found in areas that
border the _____ Sea. Summers are hot and _____ .
In summer _____ is a major problem. The _____ ____
_____ is one area that has this climate. The most common tree is
the _____ tree. Other tree crops include _____ and
_____.

12 The chart below shows the temperature and rainfall levels for a
    typical Mediterranean area.
    (a) What is the difference between the maximum and minimum
        temperatures in January? Give your answer in °C.
    (b) What is the difference between the maximum and minimum
        temperatures in July? Give your answer in °C
    (c) Calculate the average amount of precipitation/rainfall during the
        winter months. Give your answer in mm.
    (d) What is the yearly **range** of precipitation in this resort?
        Give your answer in mm.

Month	J	F	M	A	M	J	J	A	S	O	N	D
Temp (°C)	7	10	13	14	18	21	23	23	21	17	12	8
Rainfall (mm)	97	71	71	38	20	3	0	0	8	20	48	97

12 The Tundra

Key words

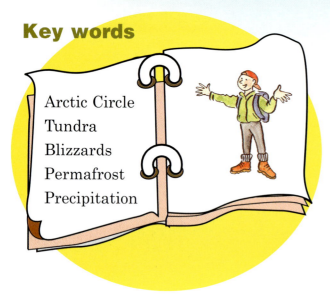

Arctic Circle
Tundra
Blizzards
Permafrost
Precipitation

When we hear about the far north of our earth we shiver. Images of icebergs and reindeer come to mind. To experience this climate you have to travel beyond the **Arctic Circle**, 66.5°N, to northern Canada, northern Europe and northern Asia. Unlike the Mediterranean type climate here, on the **tundra**, there are extremes of temperature.

Which places have a tundra type climate?

What is the climate like?

Winter

Winters are long and very cold on the tundra. This is because:

- The tundra is a long way from the equator, so these lands are very cold in winter.
- Winters in the tundra are very dark. For two months of the year, the sun doesn't rise above the horizon.
- The prevailing winds are the polar winds. These are cold winds. They come from the far north bringing cold air.

It is freezing in winter but quite warm in summer. The amount of rainfall and snowfall is low.

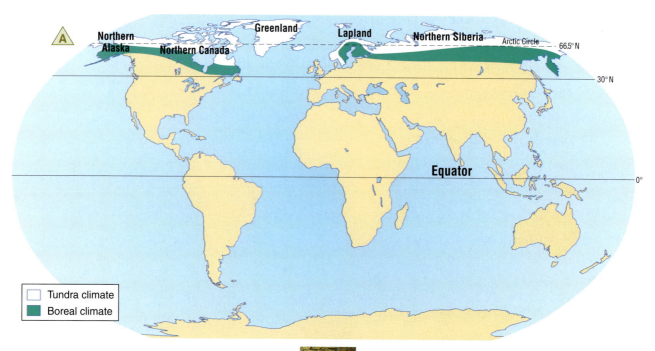

Northern Alaska · Greenland · Lapland · Northern Siberia · Arctic Circle · 66.5° N · Northern Canada · 30° N · Equator · 0°

Tundra climate
Boreal climate

The average temperature in the tundra in winter is –28°C but in the far north it can drop to –70°C. It is so cold that in places the ground is frozen hard. In fact the layer underneath the topsoil, the subsoil, is always frozen. This condition is called **permafrost.** Days are short and it can be continuously dark for two or three months.

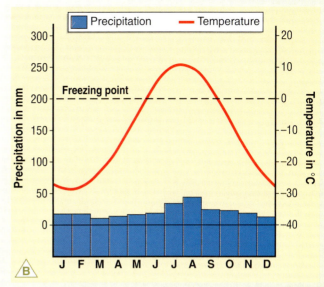

Tundra climate

Summer

Summers are short in the tundra region. Temperatures range from 3°C–16°C. In some places, north of the **Arctic Circle**, the sun hardly sets in June. This means there are 24 hours of daylight each day. That is why it is called the 'Land of the Midnight Sun'.

There is very little rain or snowfall in the tundra lands. But, because temperatures are very low, the snow that falls doesn't all melt away.

Arctic winds stir up the sea

The tundra is like a desert of snow. It is also very windy with winds reaching 48–97km per hour, so **blizzards**, or snow storms, are a danger.

Key points

- The **tundra** is a region that lies beyond the **Arctic Circle**.
- Summers are short and winters are long.
- For two months of the year there is continuous light, and for another two months there is continuous darkness.
- **Precipitation** (rainfall and snowfall) rates are low.
- Winds can be strong and **blizzards** are a danger.

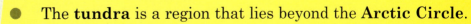

115

Natural Vegetation of the Tundra

Key words

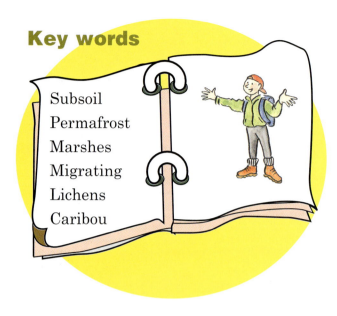

Subsoil
Permafrost
Marshes
Migrating
Lichens
Caribou

Lichens cling to rocks in the tundra

During summer, when the topsoil melts on the tundra, small plants appear. The **subsoil** (beneath) never thaws out. This layer, called **permafrost,** does not allow the water to move downwards, so during the summer thaw water stays on or near the surface. The ground can be very soggy. **Marshes**, bogs and lakes form attracting thousands of insects and **migrating** birds.

Trees are rare in the tundra. Where they do appear, in the southern parts of the region, they are small. **Lichens**, mosses and the Arctic raspberry are the most common plants.

Animal life

The arctic fox, lemmings and polar bears live there all year round. The **caribou**, a type of reindeer, live here during the summer, but migrate southwards during the winter.

The white coats of arctic foxes provide camouflage in the snowy conditions of the tundra

Case Study: The Caribou

Caribou and reindeer are types of deer. They can be almost black to brown, grey or almost white.

The caribou is the only deer in which both sexes have antlers, although those of the female are smaller.

Females love socialising and gather in herds with their young, but adult males often prefer being alone. In autumn, males fight to gather harems (groups) of about 5–40 females. The female produces one or two young after a gestation period (pregnancy) of about 240 days. Young caribou are able to run with the herd within a few hours of birth.

Some caribou migrate hundreds of miles between their breeding grounds on the tundra and winter feeding grounds farther south. Grass and other tundra plants are their main food in summer, but in winter caribou feed mainly on lichens, scraping away the snow with their hoofs to expose the plants.

Key points

- The **subsoil** in the tundra regions is always frozen. This is called **permafrost**.
- **Marshes**, bogs and lakes are common.
- **Lichens**, mosses and the Arctic raspberry are the most common plants.
- The **caribou** and reindeer survive the cold conditions.

Human Activities on the Tundra

Key words

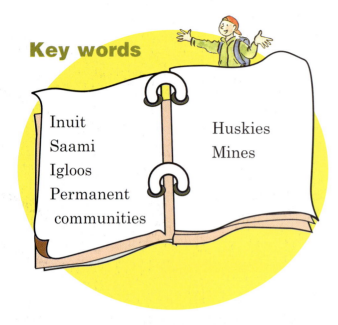

Inuit
Saami
Igloos
Permanent
communities

Huskies
Mines

Description of a typical pupil's clothes

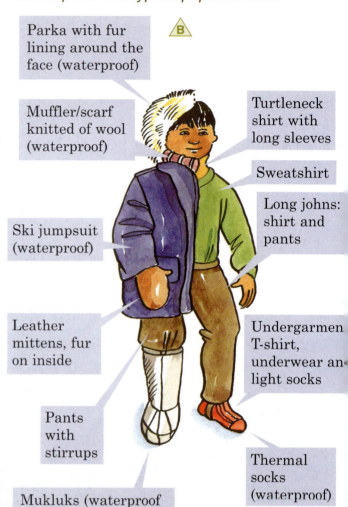

Parka with fur lining around the face (waterproof)

Muffler/scarf knitted of wool (waterproof)

Turtleneck shirt with long sleeves

Sweatshirt

Long johns: shirt and pants

Ski jumpsuit (waterproof)

Leather mittens, fur on inside

Undergarmen T-shirt, underwear an light socks

Pants with stirrups

Mukluks (waterproof boots) made of animal skin lined with fur

Thermal socks (waterproof)

The tundra is a difficult place to live and people who live here have had to adapt to these difficult conditions. The **Inuit** and **Saami** people live in the tundra regions. They used to be called Eskimos.

Inuit with his igloo in snowy conditions

Many years ago, during winter, the Inuit lived in **igloos**, which are houses made of packed snow or ice. They hunted fish and wild animals. This gave them food. It also gave them skins to keep them warm, and oil for heating and lighting. They even made their tools from the bones of animals they hunted. During summer they built animal-skin tents.

Changes in the tundra

Most of the people of the tundra lands no longer hunt and gather for a living. This has meant that many of the traditional skills handed down for generations have been lost. In the past they used all the available materials around them: snow to build their winter homes, skins to build shelters in summer, oil from seals to light their lamps, and dogs to pull their sleighs. They are no longer nomadic but have settled in **permanent communities**.

Mining

Oil

When oil was discovered, jobs and money came into the region. The government built houses, shops and schools in villages. They introduced new ways of getting from place to place. The snowmobile replaced the **huskies**, the dogs that pulled sleds.

Alaskan pipeline

Dressed for the weather

Most people now live in towns. They work in **mines** and get a regular wage. They have shops nearby. Major routeways cross these lands now.

Snowmobiles are used for transport

Difficulties with mining

- It is difficult to build roads across huge areas of 'wasteland'.
- It is hard to attract people to these isolated places, so wages must be high.
- Drilling equipment is expensive to work in such frozen seas.

Problems created by mining

- Development can disturb nature's traditions, for example, the Alaskan oil pipeline was built across a caribou migration route.

- Communities have been broken up and old traditions lost.
- Some people have found it difficult to adjust to the changes and have turned to alcohol. This has brought many problems.

Key points

- The discovery of oil and gas has changed the life of the **Inuit** people in tundra lands.

REVISION EXERCISES

Write the answers in your copybook.

1   Write out the correct answer for each of the statements below:
    (a) The tundra region is found in **Northern Norway/Ireland**.
    (b) The people who live in Northern Norway mine
        diamonds/oil and gas.
    (c) The most common animal in Northern Norway is the
        horse/caribou.

2   In your copybook match the letter in column X with the number of
    its pair in column Y.

X	Y	ANSWER
A Plant	1 Caribou	A =
B Permafrost	2 Lichens	B =
C Animal	3 Frozen subsoil	C =
D Inuit	4 People of the tundra	D =

3   Write out the following paragraph in your copybook. Use the words
    given in the box to fill in the blanks.

> **long   frozen   permafrost   short   Midnight Sun
> blizzards   caribou   Arctic Circle**

The tundra region lies to the north of the _____ _____.
Summers are _____ and temperatures rarely rise above 15°C.
Winters are _____. The strong Arctic winds that blow in these
regions stir up the snow causing _____. These cold
conditions mean that the subsoil is always _____. This is known
as _____. In summer there is continuous daylight. That is
why this area is known as the land of the _____ _____. One of
the most common animals herded in these lands is the _____.

4   Explain why the topsoil of the tundra becomes so soggy in summer.
    Use the word 'permafrost' in your answer.

5 Describe the tundra region under the following headings:
   ● Climate
   ● Vegetation
   ● Animal life

6 Examine the chart opposite and answer the
   following questions.
   (a) Name the month with the highest temperature.
   (b) Calculate the temperature range.
   (c) Name the month with the highest rainfall.
   (d) Calculate the total annual rainfall.

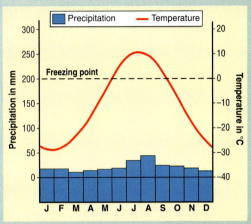

Tundra climate

7 Name three countries that have areas with the
   tundra climate.

8 Why is the tundra called the 'Land of the Midnight Sun'?

9 Describe one animal that has adapted to conditions in the tundra.

10 The native people of the tundra are said to be finding it hard to
   adapt to the changes introduced in recent times. In the case of
   the Inuit people describe two of these changes.

11 Climate influences the way we dress. Describe the typical clothes of
   a student living in a tundra region.

12 Draw a line chart using the following information on temperature in
   tundra lands.

   Jan   = −40°C
   Feb   = −37°C
   Mar   = −27°C
   April = −17°C
   May   =   3°C
   June =   9°C
   July = 12°C
   Aug   =   7°C
   Sept = −1°C
   Oct   = −7°C
   Nov   = −34°C
   Dec   = −35°C

13 Soils

Key words

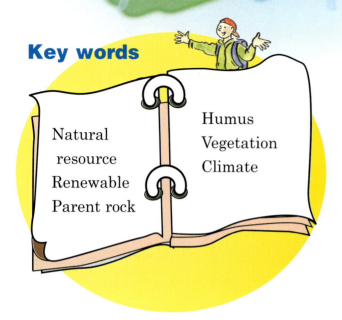

Natural resource
Renewable
Parent rock

Humus
Vegetation
Climate

A It's mucky out there today. I'm filthy.

Soil is essential for plants and animal welfare.

Soil – two different views!

What is soil?

Soil is the thin layer of loose material in which plants and crops grow. The forces of weathering and erosion break up rocks into smaller pieces. These rock pieces mix with plants, water and air to make soil.

Soil is a vital **natural resource** without which we would not be able to survive. The term **natural resource** means:

- Nature has supplied it to us – soil is **natural**.
- Soil is a **resource** – it is useful to humans.

Soil is a **renewable** natural resource. Renewable means it can be used over and over again. Sometimes people do not use the soil with enough care. In these cases it can become useless.

It can take up to 400 years to make 1 cm of soil. It takes far less time to destroy it.

The make up of soil

Soil is made up of:

- Small pieces of rock: this comes from the underlying rock, called **parent rock,** which was broken-up (weathered)
- **Humus**: this is dead plant and animal pieces in the soil
- Earthworms and other insects
- Water
- Air
- Micro-organisms: tiny creatures that help to break down the decaying plant and animal material.

Soils are made up of different ingredients

Different soils

Soils are not the same everywhere. Some soils are very fertile and crops grow well in them. In other places, the soil can be infertile and will not be good for growing crops. The type of soil in a place depends on the **climate**, **vegetation** and **parent rock.** The type of soil also depends on the influence of *people*.

We will look briefly at each of these in the following diagram. Think of how the ingredients in a cake are mixed. This helps to see how soils can differ from each other. Like cakes, it depends on the mixture.

Parent rock	Climate	Vegetation	People
Different rocks are made up of different minerals. Some minerals make fertile soils; other minerals make less fertile soils.	Frost action breaks up rocks (remember weathering). This supplies the small pieces of rock to the soil. Some rocks are broken up more easily than others. The rain supplies the water for soil. The rain can dissolve minerals in the soil and wash them downwards. If there is a lot of water in soil, it can mean there is not enough space for air to flow. This in turn will affect the type of soil.	These are the plants or crops that grow in the soil. Vegetation is influenced by climate and the type of rock in the soil. When plants die they drop leaves or roots in the soil. This plant 'litter' rots to form **humus**. This has an influence on the soil. A lot of humus usually means a fertile soil.	People can change a soil to make it support more crops. They may drain the soil or irrigate it. They may decide to add manure or chemicals to make soil fertile. These activities will influence the type of soil.

The mixture of all of these factors makes different soils.

Terra rossa (iron rich soil)

Podzol (grey soil – minerals are washed out by heavy rainfall)

Brown earth soil (rich with humus)

Key points

- Soil is an important **natural resource**.
- Soil is **renewable** if it is handled with care.
- The type of soil has to do with the mixture of **parent rock**, the **vegetation** and the **climate** in an area.
- People can change a soil so it can support more crops.

How a Soil Is Examined

Key words

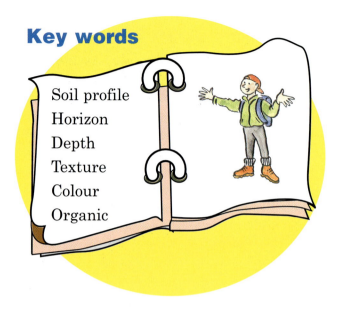

Soil profile
Horizon
Depth
Texture
Colour
Organic

Before you look at different soils in Ireland it is useful to carry out a study of the soil in your locality.

Local soil study – find out about soils in your locality

The simplest way to work out soil type is to take a **soil profile.** A profile is a side view of the layers of soil under the ground.

Local soil study

- Find a suitable place to dig, or a place where you can see a cutting of soil. The shoreline is a place where it would be possible to see such a cutting.
- With your spade, cut down into the soil until you get to the hard rock.
- Take a look at the cutting. You will notice that the soil is built up in layers. Draw the profile of these layers, labelled A, B, C.
- Look at each layer in detail. Each layer is called a **horizon**. These horizons tell you a lot about the soil in this particular area.

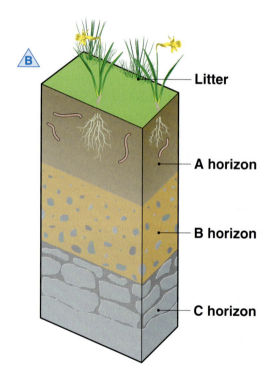

Litter

A horizon

B horizon

C horizon

Examining soil horizons

You need to examine the **depth**, **texture**, **colour** and **organic** content in each layer of your soil profile. Differences in each of these will indicate different soils.

Depth	Texture	Colour	Humus content
Depth is found by measuring the distance from the top of the soil to the hard rock below (called the parent rock!).	**Texture** is how a soil feels when you touch it. It may feel gritty or smooth. This depends on the different amounts of *sand*, *silt* and *clay* in the soil.	To describe the **colour**, first rub your fingers into the soil and then rub the soil onto white paper.	Humus content describes the amount of dead leaves, roots, plants and animals that have rotted away to form humus. The colour of the soil gives an idea of the amount of humus in the soil. If the soil is very dark in colour, it means it has lots of humus in it.

C

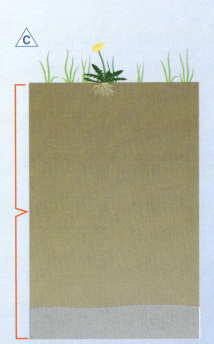

Depth

Colour

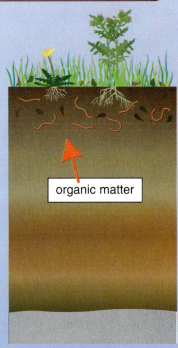

organic matter

Humus content

D

Texture

Key points

- Soils have three **horizons** (layers) – A, B and C.
- A soil is named on the basis of the **depth**, **texture**, **colour** and the amount of humus in the layers or **horizons**.

Types of soils

Key words

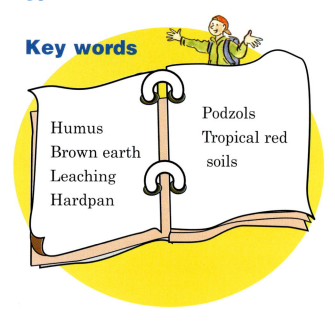

Humus
Brown earth
Leaching
Hardpan

Podzols
Tropical red
soils

There are two activities that are important in the making of different soils. These are:

● The rotting of roots and leaves in the soil.

● The action of water as it passes down through the soil.

The making of humus

Every year dead or decaying plant litter piles up on top of the soil. Think of the autumn and how the leaves cover the ground!

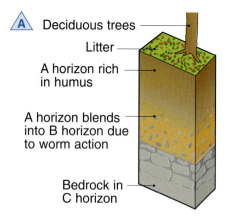

A Deciduous trees
Litter
A horizon rich in humus
A horizon blends into B horizon due to worm action
Bedrock in C horizon

Micro-organisms get to work on this plant matter – leaves and grasses – and change it into a black, jelly-like substance called **humus**. Humus is really good for soil. It has lots of dissolved minerals from the dead plant and animal litter, making it very good for growing crops.

Brown earth soils

Soils that have a lot of humus in them are dark brown in colour and are called **brown earth** soils. They are fertile soils.

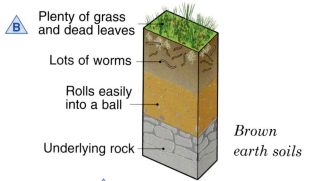

B Plenty of grass and dead leaves
Lots of worms
Rolls easily into a ball
Underlying rock

Brown earth soils

Look at map D that shows where you would find brown earth soils in Ireland.

Leaching

Leaching means the washing down of minerals through the soil. This happens when there is a lot of rain.

When rainwater falls on the soil it dissolves some of the minerals and plant litter in the soil. These minerals are then washed down through the soil where they form a **hardpan**.

The hardpan is an impermeable layer of hardened minerals in the soil. Water cannot pass down through this layer. The water builds up on top of it where it soaks the soil. The soil becomes waterlogged (saturated with water). Humus cannot form here.

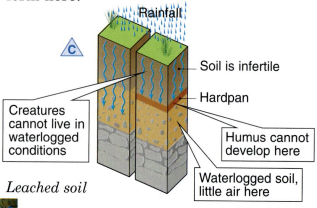

Rainfall
C
Soil is infertile
Hardpan
Creatures cannot live in waterlogged conditions
Humus cannot develop here
Waterlogged soil, little air here

Leached soil

Podzols

Soils that are heavily leached are called **podzols**. They are pale in colour and have very little humus. They are infertile. Look at map 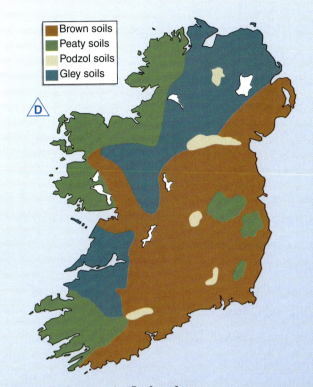 to see where you would find these soils

Main soil types in Ireland

Tropical red soils

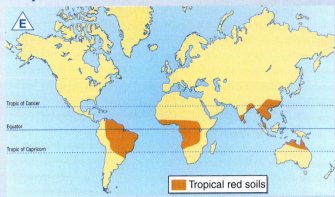

Tropical red soils are found in places close to the equator where conditions are very hot and wet. These soils are:

- Very deep
- Have lots of humus (from the forests that grow in them)
- Are rich in the mineral iron
- Red in colour (the iron adds this colour to the soil)
- Leached due to high rainfall.

This soil is a good example of how the elements of climate and vegetation mix to give a soil that is very different to any soil in Ireland. The very high temperatures and high rainfall support a thick covering of forest. In the hot, damp conditions, the humus from the trees decays very rapidly. This humus adds acid to the soil which chemically weathers the underlying rock. All of these processes create a very deep soil.

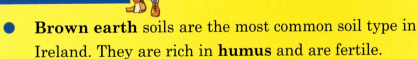

Key points

- **Brown earth** soils are the most common soil type in Ireland. They are rich in **humus** and are fertile.
- **Leaching** is the washing down of minerals in the soil.
- **Podzols** are soils that have been leached and are infertile.
- **Tropical red soils** are rich in iron and are found in places close to the equator.

Soil Erosion

Key words

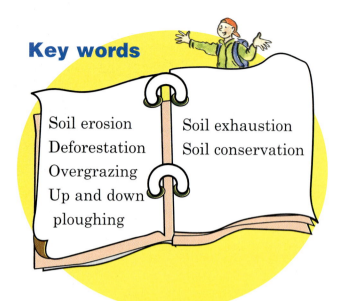

Soil erosion
Deforestation
Overgrazing
Up and down
  ploughing

Soil exhaustion
Soil conservation

Soil is an essential natural resource. We depend on soil to grow plants and crops. It takes a long time to make soil, but it can be destroyed very easily. Each year, 75 million tonnes of soil are eroded throughout the world. When soil is eroded it means that crops cannot be grown and people go hungry.

Some **soil erosion** happens naturally due to the action of rivers and wind. But soil erosion is mainly caused by human activity, when people don't handle soil properly. Soil erosion is caused by:

- **Deforestation**
- **Overgrazing**
- **Up and down ploughing** on a hill
- **Soil exhaustion**.

Deforestation

Deforestation means to cut down trees. This is usually done to clear land for farming. Plants and trees hold the soil together. When trees are cut down the wind can blow away the soil or the rain can wash it away.

Overgrazing

When too many animals are allowed to graze on a piece of land, they eat away all the vegetation. This has the same effect as deforestation. Without the grasses to bind or hold the soil together, it is more easily washed or blown away.

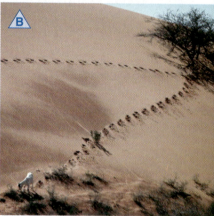

Soil exhaustion

Like a human being, soil can become tired if it is over worked! Growing too many crops on the land and not giving it time to rest can cause the soil to breakdown. It will not be able to support crops. It will dry out and the wind can blow it away. This is called **soil exhaustion**.

Up and down ploughing on a hill

When a farmer ploughs the land up and down the hill rather than around it, water can run down the furrows taking soil away with it.

Case Study of Soil Erosion: Nepal

Nepal is a country in the continent of Asia. The world's highest mountains, the Himalayas, are found here.

Nepal gets heavy rain and has steep slopes. It always had a problem with soil erosion. Lately though it has become very serious.

Nepal's population has doubled in the past 30 years and the number of tourists has increased greatly in recent times. Trees have been cut down to make more land on which to grow crops to feed these extra people. The soil has not been allowed to rest. This has led to soil erosion on the hillside.

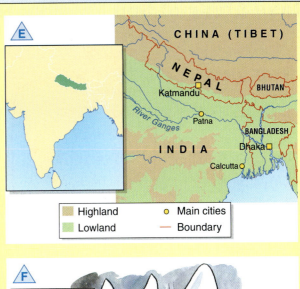

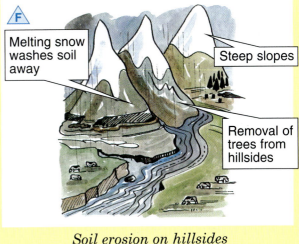

Soil erosion on hillsides

Solutions to the problem of soil erosion

You have learnt that soil is a valuable natural resource, essential for our survival. There are ways in which we can take care of soil. There are solutions to the problem. We call these solutions **soil conservation** methods.

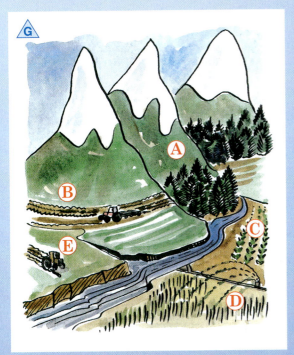

A   *Tree planting* – helps to hold the soil together.

B   *Contour ploughing* – means ploughing around a hill rather than up and down.

C   *Crop rotating* – this means changing the crops in the fields every few years. This rests the soil and can put nutrients back into it.

D   *Irrigation* – adding water helps to bind the soil particles together.

E   *Adding fertilisers to the soil* – this stops the soil from breaking-up.

Key points

- Human activities have increased the amount of **soil erosion**.
- **Soil conservation** methods help to prevent soil erosion.

Case Study of Soil Erosion: Desertification

Key words

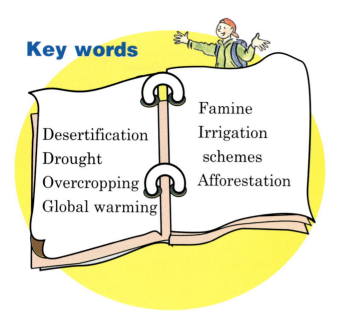

Desertification
Drought
Overcropping
Global warming

Famine
Irrigation
schemes
Afforestation

In many parts of the world the land is drying out and being blown away. It is becoming useless for farming. This is called **desertification**.

Causes of desertification

The Sahel is an area in North Africa that borders the desert. The Sahel is turning to desert because of:

● Bad farming practices.
● **Drought** (long periods without rain).

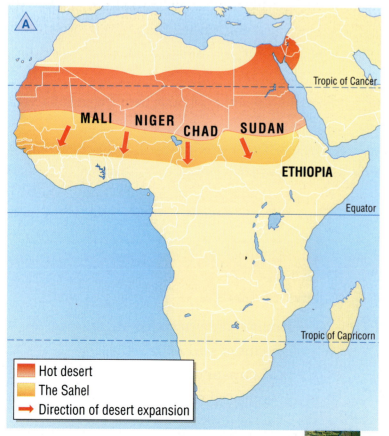

Hot desert
The Sahel
➡ Direction of desert expansion

Bad farming practices

The number of people living in the Sahel is growing. Improvements in healthcare mean people live longer. More people mean more mouths to feed. Trees are cut down to clear the land for extra crops. More animals are reared. The land is not allowed to rest because more food must be provided. All of this leads to soil erosion.

Deforestation, overgrazing by animals and **overcropping** increases the rate at which soil is eroded. When soil is eroded, crops or grasses cannot grow. The place becomes a desert.

Drought

A second reason for the growth in desert land is **drought**. This is a shortage of rainfall. While the areas close to the deserts always had an unreliable rainfall, the soil could cope with the shortage of rain for a while. But in recent years, the period between rainfalls – known as a drought period – has been getting longer. The soil has dried out completely. The wind has eroded it.

Some scientists believe that there are longer drought periods because of global warming. **Global** (worldwide) **warming** is caused by human activity, which is causing the problem of soil erosion. You will learn about this in Groundwork 2.

Results of desertification

When the soil dries out, like the soils of the deserts, crops cannot grow. There is a **famine**. People starve. Some people leave their land and go in search of a new life in another place. This leads to overcrowding in these places and again the soil is under pressure. The land is overgrazed and overcropped. The desert is spreading.

The spread of the desert

Spread of the Deserts

Africa's soil is, in general, made up of sand and laterite (a soil that is rich in the minerals iron and aluminium). This is eroded more easily and holds less water than the clay and humus-rich soils in other regions. The high iron and aluminium content makes it turn hard under the sun's heat. When the trees are cut down, the soil bakes into a concrete texture. This is impossible to farm. The soil then cannot take in rain.

But all of these factors are worsened by overcropping, overgrazing and deforestation. The problem hits the poorest farmers worst. The best land is taken for growing crops to sell abroad. Peasant farmers growing crops for local use are forced to work less fertile land. They work it too hard as they struggle to support their families. In time this soil becomes worn out. The wind then blows this away. According to UN estimates one billion tonnes of topsoil are let loose on the Ethiopian highlands each year. As a result the Sahara desert is said to be moving southwards at a rate of about 5km a year.

Solutions to the problem of desertification

To prevent the spread of the deserts, soil erosion must be stopped.

- **Irrigation schemes**: This artificial watering of the land would encourage plants to grow and would hold the soil together.
- **Afforestation schemes**: Planting trees and shrubs would hold the soil together.
- **Limit animal numbers**: Vegetation that would bind the soil would then be able to grow.
- **Choose drought resistant crops**: Crops with long roots could tap into water that has soaked deep into the soil.

Key points

- Deserts are spreading each year. This is called **desertification**.
- The Sahel region of Africa is one area into which the deserts are spreading.
- **Deforestation**, **overcropping** and overgrazing of land causes the desert land to spread.
- Longer **drought** periods are causing the deserts to spread.
- To prevent the spread of the deserts, soil erosion must be stopped.

REVISION EXERCISES

Write the answers in your copybook.

1   Leaching in soil means:
   ● Evaporating upwards of minerals
   ● Washing down of minerals by water
   ● The adding of minerals to the soil
   ● Bacteria breaking down dead vegetation

2   Irrigation means:
   ● Bringing water to dry land
   ● Draining water from wet land
   ● Protecting coastlines
   ● Using rivers for transport

3   Water is important in the soil because it helps:
   ● To dissolve minerals in the soil
   ● To stop leaching
   ● To cause desertification
   ● To drain quickly through impermeable soils

4   The Sahel is an area in which continent?
   ● Asia
   ● Europe
   ● Africa
   ● North America

5   In which soil horizon would you find humus?

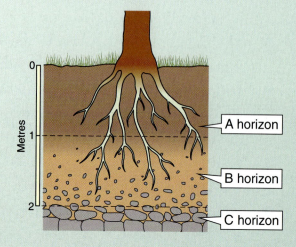

6 In your copybook match column X with the number of its pair in column Y.

X	Y	ANSWER
A Brown earths	1 Co. Kerry	A =
B Podzols	2 Co. Meath	B =
C Peat soils	3 Co. Mayo	C =
D Brown earths	4 Co. Dublin	D -

7 Name one Irish soil type that you have studied.

Draw a diagram to show the horizons (layers) in this soil type.

Label your diagram carefully.

8 Use the words provided to complete the sentences below.

> **brown   earth   soils   hard pan   podzols**

The soil types found in damp upland areas are _____

The most common soil type in Ireland is _____

The hard layer of leached material is called

_____

A soil horizon is _____

Parent rock refers to _____

9 Explain how human activities have influenced the drought in Africa. Use an example from one area.

10 Draw a map of Africa in your copy book and mark in the countries of the Sahel.

11 Soil erosion is a major problem in heavily peopled countries like India. If you were a government minister visiting an area what two solutions to the problem would you suggest? Give details and reasons for your choices.

12 Name the five ingredients from which soil is made.

13 Explain the following terms:
(a) Contour ploughing
(b) Crop rotating
(c) Soil conservation
(d) Natural resource

Key words

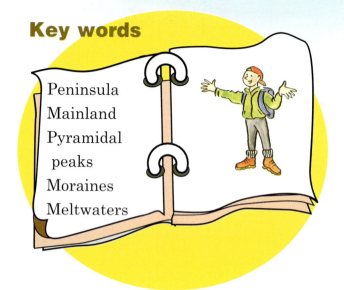

Peninsula
Mainland
Pyramidal peaks
Moraines
Meltwaters

Facts about Italy

Population: 57.8 million

Size: About four times that of Ireland

Capital: Rome

Other main cities: Milan, Turin, Genoa, Naples

Famous sites: Colosseum, Vatican City, Vesuvius, Mt Etna, Alps

Famous products: Pasta, cheeses, olive oil, cars, clothes

Member of the EU: Yes

Currency: Euro

Many of the topics you studied in this book can be explored in more detail by looking at Italy.

Italy is a long narrow country in southern Europe. Its shape is like that of a boot! It is called a **peninsula** because it is attached to the **mainland** on one end and surrounded by water on the other three sides. Look at map **A**. Italy includes the islands of Sicily, Sardinia, Capri and Elba. Find these on the map.

Italy

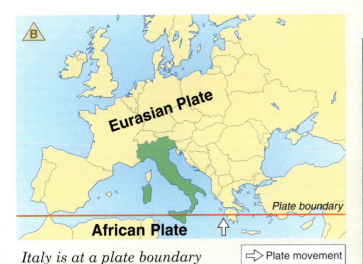

Italy is at a plate boundary

In this section, we will look at landscape features dealt with in earlier chapters.

Volcanoes – Mount Vesuvius

Mt Vesuvius is one of the most famous volcanoes in the world. It is situated in the Bay of Naples. Its wide crater is visited by many people from all over the world. It is close to the ruins of Pompeii. Pompeii was buried when Vesuvius erupted in AD 79.

Fold mountains

These fold mountains resulted from the collision of the African Plate with the Eurasian Plate. They stretch across many countries in Europe. They are made of sedimentary rocks.

Glaciation

Pyramidal peaks are dotted throughout the Alps. The beautiful lakes, Como, Garda and Maggiore, were all formed when terminal **moraines** were deposited. These blocked the movement of the **meltwaters** and formed lakes.

Key points

- Italy lies to the south of Europe and faces the Mediterranean Sea.

Earthquakes

This map of Italy shows the sites where volcanoes and earthquakes have occurred

Milan
Venice
1976
ALPS
ALPS
Po
Po
APENNINES
Tiber
Adriatic Sea
1930
Rome
2002
San Giuliano di Puglia
1983
Naples
Vesuvius
1980
Sardinia
Tyrrhenian Sea
Ionian Sea
Mediterranean Sea
1908
1955
1920
Etna
Sicily

▲ Volcanic sites
◎ Earthquake sites

Recent Italian earthquakes

1997: 13 people die and 40,000 homeless. Roof of Assisi Basilica collapses

1980: 2,500 killed and 7,500 injured in Naples

2002: over 20 children and two teachers killed in San Giuliano di Puglia. Many injured. Measured 6.4 on the Richter scale.

Mass movements

Every year in the Alps avalanches present a danger. On 18 December 2000, ten people lost their lives on a peak near Bergamo.

Floodplains

The Po Valley, just south of the Dolomite mountains, is the largest, low-lying area of southern Europe. The picture above shows the floodplain of the river. Its fertile, alluvial soils produce wheat (pasta), maize and a range of vegetables.

Delta

Delta

The River Po reaches its mouth near the Adriatic Sea. Its load is too big for the tides and currents to carry it away into the sea. It builds up as a delta.

Beaches

The Italian Riviera is a favourite area for Italians. It also attracts tourists from all over the world.

Irrigation schemes

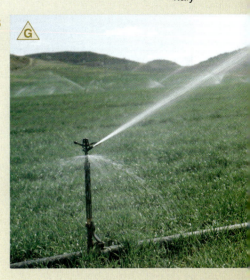

Irrigation sprinklers are now common on highly worked farms. Many have been bought with grants from the EU. This allows farmers to grow a variety of fruits and vegetables.

Soil erosion

In the Apennines, lack of money, drought, deforestation and overgrazing by animals over a long time caused the soil to be removed. This makes it difficult for crops to grow.

Mediterranean climate

The landscape of the South of Italy is dotted with olive trees and vines. Nets remain on the ground ready for the harvesting season. Sheep and goats roam the mountainsides.

Key points

- The landscape and life in Italy shows the influence of a range of physical processes.

REVISION EXERCISES

Write the answers in your copybook.

1   Italy is located in:
   ● Central Europe
   ● Southern Europe
   ● Northern Europe
   ● Eastern Europe

2   Which of the following islands belong to Italy?
   ● Malta
   ● Cyprus
   ● Sicily
   ● Corsica

3   Write the correct answer for each of the statements below:
   (a) Italy has a Mediterranean **climate/hot desert climate**.
   (b) The capital of Italy is **Milan/Rome**.
   (c) The most famous volcano in Italy is **Mauna Loa/Mt Vesuvius**.

4   Describe one earthquake that has happened in Italy. Use terms like Richter scale and epicentre in your answer.

5   Study the map and then complete the following sentences in your copybook.

Italy is bordered by France, S_____, Austria and S_____. The _____, found in the North are fold mountains. There is an extensive floodplain on the banks of the River _____. The centre of fashion is in the nearby city of M_____. Much of Italy has a _____ climate. Italy is a member of the _____.

138

6 Draw a map of Italy and mark in the following:
- The Alps and the Apennine mountains
- The River Po
- The cities of Rome, Milan, Turin, Genoa, Bologna, Naples, Palermo and Catania
- Mt Vesuvius
- The Mediterranean, Adriatic and the Ligurian Seas

7 Write a paragraph explaining why geography students would find it worthwhile to visit Italy. Give named examples.

8 Briefly explain:
- Irrigation schemes in Italy
- Soil erosion in the south of Italy

9 In your copybook create a collage of Italy that might include images of the following:
- The Italian flag
- Mt Vesuvius
- The Alps
- Beaches
- Water sprinklers
- A bag of pasta
- Red wine
- Suncream
- Parmesan, mozzarella and gorgonzola cheeses

10 Look at the photograph of the Bay of Naples.

(a) Explain, using a diagram, how this bay was formed.

(b) Explain why people spend their leisure time in bay areas.

Maps

Physical Map of the World

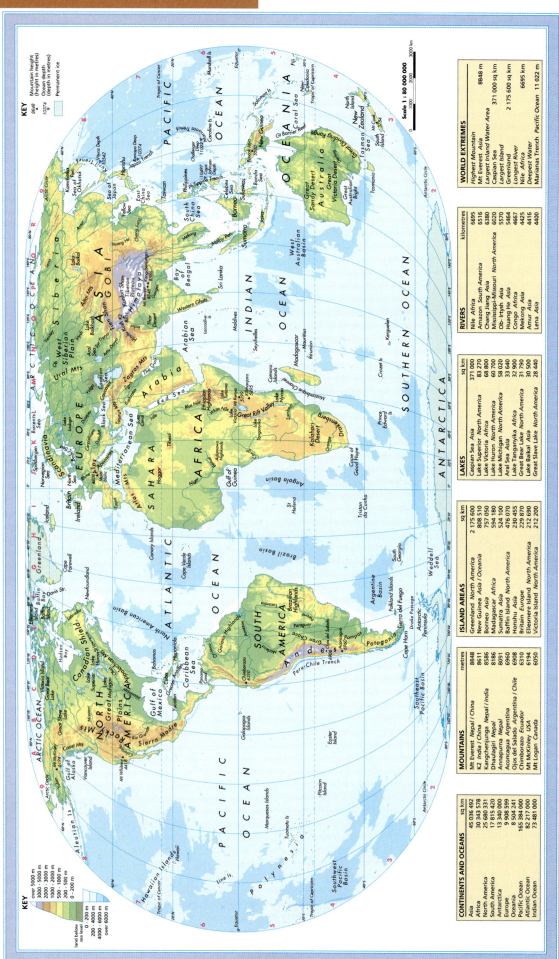

KEY

Mountain height (height in metres) — 8848
Ocean depth (depth in metres) — 10374
Permanent ice

land height
- over 5000 m
- 3000 - 5000 m
- 2000 - 3000 m
- 1000 - 2000 m
- 500 - 1000 m
- 200 - 500 m
- 0 - 200 m

land below sea level
- 0 - 200 m
- 200 - 4000 m
- 4000 - 6000 m
- over 6000 m

Scale 1 : 80 000 000
0 1000 2000 3000 km

Eckert IV projection

CONTINENTS AND OCEANS	sq km
Asia	45 036 492
Africa	30 343 578
North America	25 680 331
South America	17 815 420
Antarctica	13 340 000
Europe	9 908 599
Oceania	8 504 241
Pacific Ocean	165 384 000
Atlantic Ocean	82 217 000
Indian Ocean	73 481 000

MOUNTAINS		metres
Mt Everest	Nepal / China	8848
K2	India / China	8611
Kangchenjunga	Nepal / India	8586
Dhaulagiri	Nepal	8186
Annapurna	Nepal	8091
Aconcagua	Argentina	6960
Ojos del Salado	Argentina / Chile	6908
Chimborazo	Ecuador	6310
Mt McKinley	USA	6194
Mt Logan	Canada	6050

ISLAND AREAS		sq km
Greenland	North America	2 175 600
New Guinea	Asia / Oceania	808 510
Borneo	Asia	757 050
Madagascar	Africa	594 180
Sumatra	Asia	524 100
Baffin Island	North America	476 070
Honshu	Asia	230 455
Britain	Europe	229 870
Ellesmere Island	North America	212 690
Victoria Island	North America	212 200

LAKES		sq km
Caspian Sea	Asia	371 000
Lake Superior	North America	83 270
Lake Victoria	Africa	68 800
Lake Huron	North America	60 700
Lake Michigan	North America	58 020
Aral Sea	Asia	33 640
Lake Tanganyika	Africa	32 900
Great Bear Lake	North America	31 790
Lake Baikal	Asia	30 500
Great Slave Lake	North America	28 440

RIVERS		kilometres
Nile	Africa	6695
Amazon	South America	6516
Chang Jiang	Asia	6380
Mississippi-Missouri	North America	6020
Ob-Irtysh	Asia	5570
Huang He	Asia	5464
Congo	Africa	4667
Mekong	Asia	4425
Amur	Asia	4416
Lena	Asia	4400

WORLD EXTREMES		
Highest Mountain	Mt Everest Asia	8848 m
Largest Inland Water Area	Caspian Sea	371 000 sq km
Largest Island	Greenland	2 175 600 sq km
Longest River	Nile Africa	6695 km
Deepest Water	Marianas Trench Pacific Ocean	11 022 m

Political Map of the World

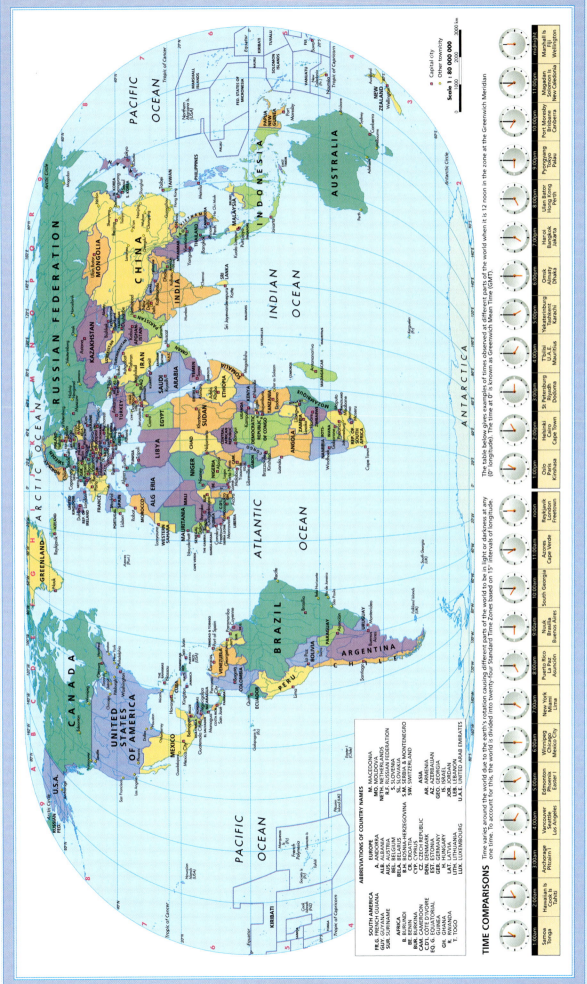

Eckert IV projection

Scale 1 : 80 000 000

■ Capital city
○ Other town/city

ABBREVIATIONS OF COUNTRY NAMES

SOUTH AMERICA
FR.G. FRENCH GUIANA
GUY. GUYANA
SUR. SURINAME

AFRICA
B. BURUNDI
BE. BENIN
BUR. BURKINA
CAM. CAMEROON
CDI CÔTE D'IVOIRE
EQ. G. EQUATORIAL
GUINEA
GH. GHANA
R. RWANDA
T. TOGO

EUROPE
A. ANDORRA
ALB. ALBANIA
AUS. AUSTRIA
BEL. BELGIUM
BELA. BELARUS
B.H. BOSNIA-HERZEGOVINA
CR. CROATIA
CYP. CYPRUS
CZ. CZECH REPUBLIC
DEN. DENMARK
EST. ESTONIA
GEO. GEORGIA
GER. GERMANY
H. HUNGARY
LAT. LATVIA
LITH. LITHUANIA
LUX. LUXEMBOURG

M. MACEDONIA
MO. MOLDOVA
NETH. NETHERLANDS
R.F. RUSSIAN FEDERATION
SL. SLOVENIA
SLK. SLOVAKIA
S.M. SERBIA & MONTENEGRO
SW. SWITZERLAND

ASIA
AR. ARMENIA
AZ. AZERBAIJAN
GEO. GEORGIA
IS. ISRAEL
JOR. JORDAN
LEB. LEBANON
U.A.E. UNITED ARAB EMIRATES

TIME COMPARISONS

Time varies around the world due to the earth's rotation causing different parts of the world to be in light or darkness at any one time. To account for this, the world is divided into twenty-four Standard Time Zones based on 15° intervals of longitude.

The table below gives examples of times observed at different parts of the world when it is 12 noon in the zone at the Greenwich Meridian (0° longitude). The time at 0° is known as Greenwich Mean Time (GMT).

1:00am Samoa Tonga	2:00am Hawaiian Is Cook Is Tahiti	3:00am Anchorage Pitcairn I	4:00am Vancouver Seattle Los Angeles	5:00am Edmonton Phoenix Easter I	6:00am Winnipeg Chicago Mexico City	7:00am New York Miami Lima	8:00am Puerto Rico La Paz Asunción	9:00am Nuuk Brasília Buenos Aires	10:00am South Georgia	11:00am Azores Cape Verde	noon Reykjavik London Freetown	1:00pm Oslo Paris Kinshasa	2:00pm Helsinki Cairo Cape Town	3:00pm St Petersburg Riyadh Dodoma	4:00pm T'bilisi U.A.E. Mauritius	5:00pm Yekaterinburg Tashkent Karachi	6:00pm Omsk Almaty Dhaka	7:00pm Hanoi Bangkok Jakarta	8:00pm Ulan Bator Hong Kong Perth	9:00pm Pyongyang Tokyo Palau	10:00pm Port Moresby Brisbane Canberra	11:00pm Magadan Solomon Is New Caledonia	midnight Marshall Is Fiji Wellington

World Map of Plates – Location of Earthquakes and Volcanoes

Eckert IV projection

Scale 1 : 85 000 000

4000 3000 2000 1000 0 1000 2000 3000 4000 km

Map labels

PACIFIC PLATE

PHILIPPINE PLATE

INDO-AUSTRALIAN PLATE

EURASIAN PLATE

ARABIAN PLATE

AFRICAN PLATE

SOMALI PLATE

ANTARCTIC PLATE

NORTH AMERICAN PLATE

SOUTH AMERICAN PLATE

CARIBBEAN PLATE

COCOS PLATE

NAZCA PLATE

JUAN DE FUCA PLATE

ASIA

EUROPE

AFRICA

AUSTRALIA

ANTARCTICA

NORTH AMERICA

SOUTH AMERICA

PACIFIC OCEAN

ATLANTIC OCEAN

INDIAN OCEAN

Tropic of Cancer
Equator
Tropic of Capricorn
Arctic Circle
Antarctic Circle

Greenland, Britain, Ireland, Iceland, Hekla, Japan, Unzen, Oyama, Pinatubo, Philippine Fault, Borneo, Sumatra, Java, Galunggung, New Guinea, Rabaul, New Zealand, African Rift System, Lake Nyos, Hawaiian Islands, Kilauea, Mt St Helens, San Andreas Fault, El Chichon, Soufrière Hills, Nevado del Ruiz, Galeras, El Llaima

CRUSTAL PLATES

The earth is made up of three main layers. The outer layer, known as the crust, ranges in thickness from a few kilometres under the oceans to almost 50 km under mountain ranges. The middle layer, known as the mantle, makes up 82% of the earth's volume. At the centre (core) of the earth, temperatures reach 4300 °C.

Crust 6-50km
Upper Mantle (solid) 370km
Transitional Zone 600km
Lower Mantle (solid) 1700km
Outer Core (liquid) 2100km
Inner Core (solid) 1350km

—— Plate boundary

EARTHQUAKES

Earthquakes occur most frequently along the junction of the plates which make up the earth's crust. They are caused by the release of stress which builds up at the plate edges. When shock waves from these movements reach the surface they are felt as earthquakes which may result in severe damage to property or loss of lives.

● High magnitude earthquake (over 7.8 on Richter scale)

Year	Location	Force	Deaths
1990	Northwestern Iran	7.7	50 000
1990	Luzon *Philippines*	7.7	1600
1991	Georgia	7.1	114
1991	Uttar Pradesh *India*	6.1	1600
1992	Flores *Indonesia*	7.5	2500
1992	Erzincan *Turkey*	6.8	500
1992	Cairo *Egypt*	5.9	550
1993	Northern *Japan*	7.8	185
1993	Maharashtra *India*	6.4	9700
1994	Kuril Islands *Japan*	8.3	10
1995	Kobe *Japan*	7.2	5200
1995	Sakhalin *Russian Fed*	7.6	2500
1996	Yunan *China*	7.0	251
1997	Quae'n *Iran*	7.1	2400
1999	Izmit *Turkey*	7.4	*15 000

* estimate

VOLCANOES

The greatest number of volcanoes are located in the 'Ring of Fire' around the Pacific Ocean. Violent eruptions often occur when two plates collide and the heat generated forces molten rock (magma) upwards through weaknesses in the earth's crust. Thousands of volcanic eruptions of varying intensity occur each year.

▲ Active volcano

Year	Location
1980	Mt St Helens *USA*
1981	Hekla *Iceland*
1982	El Chichon *Mexico*
1982	Galunggung *Indonesia*
1983	Kilauea *Hawaii*
1983	Oyama *Japan*
1985	Nevado del Ruiz *Colombia*
1986	Lake Nyos *Cameroon*
1991	Pinatubo *Philippines*
1991	Unzen *Japan*
1993	Mayon *Philippines*
1993	Galeras *Colombia*
1994	El Llaima *Chile*
1994	Rabaul *Papua New Guinea*
1997	Soufrière Hills *Montserrat*

The Physical Map of Ireland

Conic Equidistant projection

Political Map of Ireland

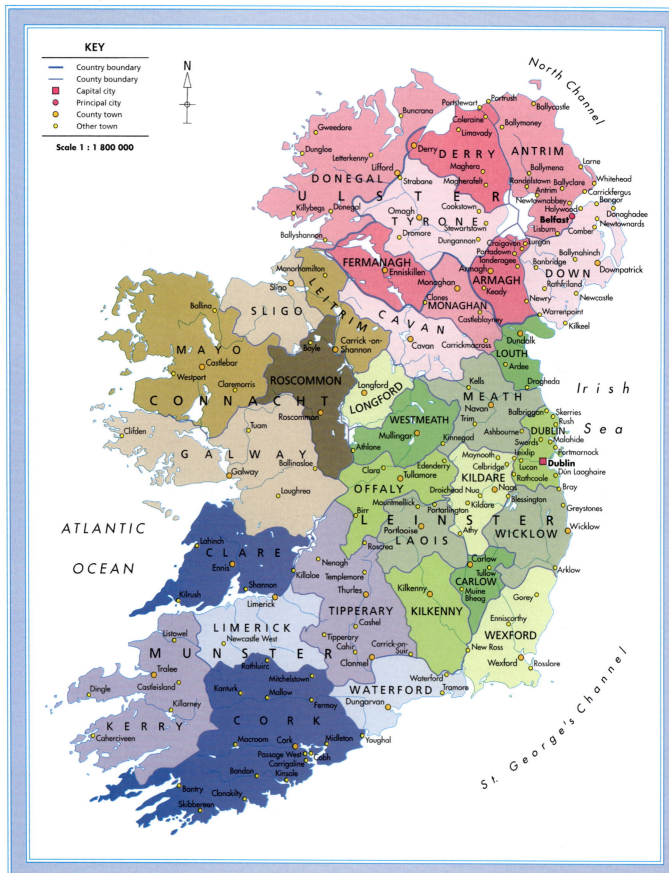

KEY

- —— Country boundary
- —— County boundary
- ■ Capital city
- ● Principal city
- ● County town
- ● Other town

Scale 1 : 1 800 000

N

North Channel

ULSTER

DONEGAL
Gweedore
Dungloe
Letterkenny
Lifford
Killybegs
Donegal
Ballyshannon

DERRY
Derry
Maghera
Magherafelt
Cookstown

ANTRIM
Portstewart
Portrush
Ballycastle
Coleraine
Ballymoney
Limavady
Ballymena
Larne
Randalstown
Ballyclare
Whitehead
Antrim
Carrickfergus
Newtownabbey
Bangor
Holywood
Donaghadee
Belfast
Newtownards
Lisburn
Comber

TYRONE
Omagh
Dromore
Stewartstown
Dungannon
Craigavon
Portadown
Lurgan
Ballynahinch
Tanderagee
Banbridge
DOWN
Downpatrick
Armagh
Rathfriland
Newcastle
ARMAGH
Keady
Newry
Warrenpoint
Kilkeel

FERMANAGH
Manorhamilton
Enniskillen
Monaghan
MONAGHAN
Clones
Castleblayney

CAVAN
Cavan
Carrickmacross

Sligo
LEITRIM
SLIGO
Ballina

MAYO
Castlebar
Westport
Claremorris

Boyle
Carrick-on-Shannon
ROSCOMMON
Roscommon

CONNACHT

Clifden

GALWAY
Tuam
Galway
Ballinasloe
Loughrea

Longford
LONGFORD
Athlone

WESTMEATH
Mullingar
Kinnegad

Clara
Tullamore
OFFALY
Birr
Mountmellick
Portlaoise
LAOIS
Roscrea

LOUTH
Dundalk
Ardee
Drogheda

MEATH
Kells
Navan
Trim
Balbriggan
Ashbourne

DUBLIN
Skerries
Rush
Swords
Malahide
Portmarnock
Dublin
Leixlip
Lucan
Dún Laoghaire
Rathcoole
Maynooth
Celbridge

KILDARE
Edenderry
Droichead Nua
Naas
Kildare
Blessington
Athy
Bray
Greystones
Wicklow
WICKLOW

LEINSTER

Carlow
Tullow
CARLOW
Muine Bheag
Gorey
Arklow

Nenagh
Killaloe
Templemore
Thurles
TIPPERARY
Cashel
Tipperary
Cahir
Carrick-on-Suir
Clonmel

Kilkenny
KILKENNY

Enniscorthy
WEXFORD
New Ross
Wexford
Rosslare

ATLANTIC OCEAN

CLARE
Lahinch
Ennis
Shannon
Kilrush
Limerick

LIMERICK
Listowel
Newcastle West
Rathluirc

MUNSTER

KERRY
Tralee
Dingle
Castleisland
Killarney
Caherciveen

CORK
Kanturk
Mitchelstown
Mallow
Fermoy
Macroom
Cork
Passage West
Cobh
Carrigaline
Bandon
Kinsale
Clonakilty
Skibbereen
Bantry
Midleton
Youghal

WATERFORD
Dungarvan
Waterford
Tramore

Irish Sea

St. George's Channel

Political Map of Europe

KEY
——	Country boundary
——	Road
——	Railway
······	Ferry route
✈	Airport
■	Capital city
●	Large town or city
●	Other town or city

Scale 1 : 20 000 000

0   200   400   600   800 km

A. ANDORRA
L. LIECHTENSTEIN
M. MONACO
S.M. SAN MARINO

Albers Equal Area Conic projection

145

Glossary

Abrasion: A process of erosion that uses rocks to wear down other rocks.

Acid rain: Rainwater with chemicals in it resulting from the burning of fossil fuels.

Active volcano: A volcano that has erupted recently and is likely to erupt again.

Alluvium: A fertile soil deposited by a river when it floods an area.

Altitude: The height of a place above sea level.

Anticyclone: A weather system with high pressure at the centre. Brings dry weather.

Ash and dust: Fine materials thrown out by a volcano.

Atmosphere: The air around the earth.

Backwash: The water that moves back into the sea from breaking waves.

Basalt: A black igneous rock that forms on the surface.

Beach: An area of sand and pebbles found between the areas reached by high tides and low tides.

Brown earth soils: Fertile soils that formed in areas that were once covered in deciduous trees.

Carbonation: A process of chemical weathering that occurs when carbon dioxide reacts with rainwater to dissolve rocks.

Clay: A finely grained substance that sticks together easily and is found in most types of soil.

Cliff: A steep slope formed when waves erode areas of a coastline.

Climate: The average weather conditions of a place.

Clints: The blocks of limestone that are separated by grikes.

Condensation: The process by which water vapour changes to liquid water when it is cooled.

Conflict: Disagreement over how a resource is used.

Coniferous trees: Trees that have shallow roots and are evergreen (have leaves all year).

Conservation: The protection and preservation of animals, plants, buildings and the environment.

Convectional rain: Rain that is produced when warm air rises.

Core: The centre of the earth.

Corrie: A hollow with three steep sides found in mountain areas that have been glaciated. It marks the birthplace of the glacier.

Crater: A roughly circular opening at the top of a volcano.

Crust: The outer rock layer of the earth. This is where we live.

Dam: A barrier that holds back water on a river.

Deciduous trees: Trees that have deep roots and lose their leaves in winter.

Deforestation: The clearing and destruction of forests. This often leads to soil erosion.

Delta: A flat area of fertile soil that forms at the mouth of a river or a lake.

Deposition: The laying down of materials carried by rivers, sea or ice.

Depression: A weather system with low pressure at its centre. Brings rain.

Dormant volcano: A volcano is known to have erupted but has not done so in recent times.

Drought: A long spell of dry weather.

Drumlins: Oval-shaped hills of boulder clay deposited by the ice.

Earthquake: A movement or vibration of the earth's crust.

Environment: The area in which people, animals and plants live.

Epicentre: The point on the earth's surface where earthquake damage is greatest.

Equator: An imaginary line around the earth halfway between the north and south poles.

Erosion: The wearing away and removal of rock and soil by rivers, sea and ice.

Erratic: A large boulder deposited by glaciers and formed of a different rock to the rock in the local area in which it is found.

Eskers: Snake-like low hills of sand and gravel formed when rivers flowed from the melting glaciers, towards the end of the Ice Age.

Evaporation: The process by which liquid water changes to water vapour when warmed.

Extinct volcano: A volcano that has not erupted in historic times and is not expected to erupt again.

Flooding: The movement of water over an area that is usually dry. It may be a river flowing over flat land beside it, or the sea covering a low-lying coastal area.

Floodplain: The flat area at the bottom of a river valley which is often flooded. It is very fertile.

Focus: The point deep in the earth's crust where the earthquake begins.

Fold mountains: Mountains whose rocks were crushed and folded as plates collided.

Fossil fuels: Fuels such as coal, oil and gas. They were formed from the remains of animals and plants millions of years ago.

Freeze-thaw action: The weathering of rocks by the action of frost.

Frontal rain: When warm air has to rise over cold air. This happens in a depression.

Granite: Light-coloured igneous rocks that form below the surface.

Gravity: A force that draws all things towards the earth's surface.

Grikes: The cracks in limestone that were once joints.

Groundwater: Fresh water that is stored in rocks and the soil. It may pass from there to the sea.

Hot deserts: Places near the equator which are very dry, with low rainfall and little vegetation.

Humus: Remains of plants and animals left in the soil.

Hydraulic action: A process of erosion that uses the power of water to wear down rocks.

Hydroelectricity: Energy got from using fast flowing water in rivers.

Igneous rocks: Rocks that are formed from molten material.

Impermeable rocks: Rocks that don't allow water to pass down through them.

Interlocking spurs: Pieces of hard rock that stick out along the course of a river in its youthful stage.

Irrigation: The artificial watering of the land in a dry climate.

Isobar: A line on a map joining places with the same pressure.

Karst: A limestone landscape that has both surface and underground features of weathering.

Landforms: Natural features formed by rivers, the sea, ice and volcanoes.

Landscape: The scenery or appearance of an area.

Latitude: The distance a place is north or south of the equator.

Lava: Molten rock flowing out of the ground, usually from a volcano.

Leaching: A process in soils, where water drains all the minerals downwards leaving a hard infertile layer in the top layer of the soil.

Levees: Natural raised banks of material deposited by rivers. If high enough they can prevent further flooding.

Limestone: A sedimentary rock formed from the remains of sea animals.

Load: The material carried by a river.

Longshore drift: The movement of material by the waves along a coastline.

Magma: Molten rock below the earth's surface.

Magma chamber: Where molten rock is found and stored deep below the earth's surface.

Mantle: The layer of the earth below the crust and above the core.

Marble: A metamorphic rock formed when limestone was changed.

Marram grass: A type of tough grass used to bind sand together on sand dunes.

Mass movement: The movement of large amounts of material down a slope under the influence of gravity.

Meander: A large bend in a river.

Mediterranean climate: Places that have hot dry summers and warm moist winter and lie 30°–40° north and south of the equator.

Mid-Atlantic Ridge: A line of volcanic mountains that runs down the middle of the Atlantic marking the place where two plates separate.

Moraine: The material that is deposited by glaciers.

Natural vegetation: Vegetation that has not been affected by human activity.

North Atlantic Drift: A warm ocean current that brings mild weather conditions to the west of Ireland.

Overgrazing: Damaging pasture (fields) by keeping too many animals on it. Can lead to soil erosion.

Permafrost: The condition where the subsoil is always frozen.

Permeable rocks: Rocks that allow water to pass down through them.

Plate boundary: The places where plates meet.

Plates: Large sections of the earth's crust.

Podzols: A type of soil that is slightly acid and was once covered with coniferous trees.

Pollution: Noise, dirt and other harmful substances produced by people and machines.

Precipitation: Water in any form which falls to earth. It includes rain, snow and hail.

Prevailing winds: The direction from which the wind usually blows.

Pressure: The weight of air pressing down on the earth's surface.

Processes: The stages or ways involved in any action of change (inputs into outputs).

Pumice: A lightweight volcanic rock that is thrown out as a volcano erupts.

Quartz: A mineral found in rocks and used in watches.

Quartzite: A metamorphic rock formed when sandstone was heated and pressurised to change.

Relief rain: Rain caused by air being forced to rise over hills and mountains.

Richter scale: A scale used to measure the strength of an earthquake.

'Ring of Fire': A circle of volcanoes around the edge of the Pacific Ocean.

River basin: The area of land drained by a river and its tributaries.

River mouth: The end of a river where it enters the sea or a lake.

River source: Where a river begins.

Sand: A gritty, fine material found in most soils.

Sandstone: A sedimentary rock formed from layers of sand deposited on the beds of seas and lakes.

Sedimentary rocks: Rocks made from the remains of dead plants, animals and broken-up pieces of other rocks and laid down in layers.

Seismograph: An instrument used to measure earthquakes.

Silt: Material deposited by a river.

Soil: The loose material on the earth's surface in which plants grow.

Soil erosion: The wearing away and loss of soil due to wind, rain, running water (rivers) and human activity.

Soil horizon: A layer of soil.

Soil profile: A side view of all the layers of soil.

Soil texture: How soil feels to the touch. It may be sticky, smooth or gritty.

Stalactites: These are carrot-like shapes that hang from the roof of a limestone cave.

Stalagmites: These are cone-shaped features that build up on the floor of a limestone cave.

Subsoil: The layer of soil found between the topsoil and the parent rock below.

Surface water: Water which lies on top of, or flows over, the ground.

Swash: The water that moves towards the shore when waves break.

Temperature: A measure of how warm or cold the air is.

Topsoil: The upper layer of soil.

Transportation: The movement of eroded material by river, sea and ice.

Tributary: A small stream or river that joins up with a larger one.

Tsunamis: Tidal waves caused when earthquakes happen in the sea.

Tundra: Places that have short, cool summers and long, cold winters and are found north of the Arctic Circle.

U-shaped valley: A glaciated valley with steep sides and a wide flat floor.

Vent: An opening in the earth's surface through which material is forced during a volcanic eruption.

Volcano: A cone-shaped mountain or hill often made from lava and ash.

V-shaped valley: The shape of a river valley when seen from the air.

Waterfall: A sudden fall of water formed from river erosion in its youthful stage.

Weather: The day to day conditions of the atmosphere. It includes temperature, rainfall and wind.

Weathering: The break-up of rock by the natural forces of weather. This produces soil.

Wind: The movement of air from areas of high pressure to areas of low pressure.